Six Scorched Roses

A CROWNS OF NYAXIA NOVELLA

CARISSA BROADBENT

Please note that this story contains subject matter that may be difficult for some, including discussion of terminal illness, death, violence, and explicit sexual situations.

PART ONE

THE FIRST ROSE

CHAPTER ONE

The first time I met death, it was in my first breaths — or rather, the first breaths I didn't take. I was born too small, too sickly, too quiet. My father used to say that he'd never heard such a silence as when I was born — several terrible minutes in which no one said a word — and that when I finally started to wail, he'd never been so grateful to hear a scream.

Death never left, though. That became clear quickly, even before anyone wanted to acknowledge it.

The truth came the second time I met death, eight years later, when my sister was born. She, unlike me, screamed from the moment she came into the world. My mother, on the other hand, went forever silent.

My father had been right. There was nothing worse than that kind of silence.

And it was in that horrible soundlessness, as I stifled my coughs and my tears with the back of my hand, that the healer gave me a strange look. Later, after my mother's funeral, he would pull me aside.

"How long has your breathing been that way?" he would ask.

Death always followed me, you see.

It quickly became clear that I wouldn't have long to live. In the beginning, they tried to hide this from me. But I'd always liked knowing things. I was bad at reading people, but I was good at understanding science. I knew death even before I could name it.

But the third time I met death, it hadn't come for me.

It was given to the town of Adcova like a silk blanket, settling slowly over our lives, placed there by one of the gods themselves.

Here's the thing about the God of Abundance. Abundance wears many faces. The god of plenty is also the god of decay. There can be no life without death, no feast without famine.

Like all the other gods, Vitarus is a fickle and emotional being. The difference between excess and absence a mere whim of his moods. Entire lives—entire towns—made or unmade by a thoughtless wave of his hand.

For a long time, Vitarus smiled upon Adcova. We were a flourishing farm town, nestled in a fertile patch of land. We worshipped all the gods of the White Pantheon, but Vitarus was the god of the farmer, and so he was our favored deity. For a long time, he treated us well.

That changed slowly, in the beginning. One spoiled crop, then two. Weeks and then months of nothing. Then, one day, it changed all at once.

You can feel it in the air when a god is nearby. I felt it that day. I opened my eyes and stared at the ceiling and could have sworn I smelled the smoke of funeral pyres.

I went outside. It was cold, my breath coming in little puffs of white. I was fifteen, but looked younger. My body shook. I was very thin, no matter how much I ate. Death stole every mouthful, you see, and it had been especially hungry lately.

To this day, I'm not sure why I went to the door. I was confused at first by what I was looking at. I thought my father was working in the fields, his form hunched and crouched in the dirt. But instead of the sea of greenery around him, there was only withered brown, coated with the wet, deadly sheen of frost.

I had never been good at seeing the things that people didn't say. But even then, as a child, I knew that my father was broken. He clutched fistfuls of dead crops in his hands, sagging over them like lost hope.

"Fa?" I called out.

He looked over his shoulder at me. I pulled my shawl tighter around myself and shivered, despite the beads of sweat on my forehead. I couldn't stop the shaking.

He looked at me the same way that he looked at those dead crops. Like I was the corpse of a dream, buried in everything he couldn't save.

"Go back inside," he said.

I almost didn't.

For years, I would wish that I hadn't.

But how was I supposed to know that my father was about to curse a god that would curse us back?

That's when the plague came. My father was the first to go. The rest, slower. Years passed, and Adcova withered like the crops in the field that morning my father had damned us all.

It's strange to watch the world wither around you. I had always put such stock in knowing things. Even the things that can't be known—the power of a god, the actions of a cruel unfair fate—have a defined edge to them, a pattern that I could pull apart.

5

I learned everything about the illness. I learned how it stole breath from lungs and blood from veins, how it reduced skin to layers and layers of fine dust until there was nothing left but rotting muscle. Yet, there was always something more there, something I couldn't ever really understand. Not truly.

So much lived in that gap—the gap between the things I knew and the things I didn't. So much died there. No matter how many medicines I brewed or remedies I tested.

The gap had teeth like the vampires across the sea. Teeth sharp enough to eat us all alive.

Five years passed, ten, fifteen. More people grew sick.

The disease came for all of us in the end.

CHAPTER TWO

I always kept my workspace clean, but I took care to make it extra organized that evening. Beneath the waning light of sunset, which splashed bloody pink over my desk, I carefully sorted my notes and instruments. Everything was in its perfect place when I was done. Even a stranger could have sat down at my table and resumed my work. I figured this was practical, just in case I didn't come back. I was expendable, but the work wasn't.

I surveyed my handiwork with a critical eye, then went out to the greenhouse. It wasn't a very pretty place—full not with colorful flowers but instead spiny leaves and vines stuffed into glass jars. Not much wanted to grow here these days. Only one little piece of beauty glinted in the back, beyond the door that led to the fields. Once, when I was very young, these fields were full of crops. Now, only one patch of dirt flourished—a cluster of rosebushes, black flowers perched upon emerald leaves, each petal outlined in a shock of red.

I carefully clipped a single flower, tucked it into my bag with special care, then went to the yard.

Mina was sitting in the sun. It was warm, but she kept a blanket over her lap anyway. She turned to me and squinted into the waning light, looking at my bag. "Where are you going?"

"Errands," I said.

She frowned. She saw through the lie.

I paused beside her for a moment—observing the darkness under her delicate fingernails, the heaviness of her breathing. Observing most of all the fine coating of flesh-colored dust that settled over the chair and her blanket. Her very skin abandoning her, as death crept closer.

I put my hand on my sister's shoulder, and for a moment I considered telling her that I loved her.

I didn't say it, of course.

If I did that, she would know where I was going and try to stop me. Besides, a word was useless compared to what I was about to do. I could show my love in medicine and math and science. I couldn't show it to her in an embrace—and what good would a thing like that do, anyway?

Besides, if I hugged her, maybe I wouldn't be able to let her go.

"Lilith—" she started.

"I'll be back soon," I said.

BY THE TIME I reached the doors, I was panting and sweating. I paused at the doorstep, taking a moment to collect myself. I didn't want whatever was about to greet me to see me looking like a mangy dog. I glanced over my shoulder, down the dozens of marble steps I had just scaled, and into the forest beyond. My town was not visible from here. It had been a long, long walk.

Next time, I'd take a horse.

I craned my neck up to the house before me. It was a strange collection of architectural elements—flying buttresses and arched windows and marble columns, all mashed together in a mansion that really should have looked ridiculous, but instead stood in stubborn and intimidating indifference.

I drew in a deep breath and let it out.

Then I knocked, and waited.

And waited.

Nothing.

After a few minutes, I knocked again, louder.

Waited.

Nothing.

I knocked a third time, a fourth. And then, finally, I thought to myself, *Well, this is the stupidest thing I'll ever do,* and tried opening the door.

The door, to my luck—or misfortune—was unlocked. The hinges squealed like this door had not been opened for a very, very long time. I had to throw myself against the mahogany to get it to budge.

It was silent within. Dusty. The interior of the house was just as strangely inconsistent in style as the exterior, though it took a few minutes for my eyes to adjust enough to see that. It was dark inside, the only light the moonlight spilling from behind me. The silver outlined the silhouettes of countless objects—sculptures and paintings and artifacts and so much more I couldn't even begin to take in. Gods, it was mesmerizing.

"Hello?" I called out.

But there was no sound. No movement, save for the faint rustling of moth-bitten gauze curtains.

Maybe he was dead. No one had seen him for a few decades. I'd be disappointed if I came all this way just to discover a rotting corpse. *Did* his kind rot? Or did they just —

"It appears," a deep voice said, "a little mouse has made its way into my home."

CHAPTER THREE

There's nothing to be afraid of, I told myself, but that did nothing to stop the hairs from rising on the back of my neck.

I turned.

And though I was expecting it, the sight of him standing on the stairwell, enveloped in shadow, still made me jump — the way one jumps when a snake moves in the underbrush beneath your feet.

It took a moment for my eyes to adjust to the deeper darkness of the stairwell. He stood at the top of the stairs, peering down at me with the vague curiosity of a hawk. He had long, dark brown hair, slightly wavy, and a neat beard. He wore a plain white shirt and black trousers, unremarkable if a little outdated. He was large, but not monstrously so. I saw no horns nor wings, no matter how hard I squinted into the dark.

I was almost a little disappointed by how... *normal* he looked.

Yet, the way he moved betrayed his inhumanity — or rather, the way he didn't. He was still the way stone was still, no minuscule shift to his muscles or rise or fall of his shoulders, no blink or waver of his gaze as it drank me in. You don't realize how much you notice those things in a person until they aren't there, and suddenly every instinct inside of you is screaming, *This is wrong!*

He approached down the stairs, the moonlight illuminating bright amber eyes and a slow smile—a smile that revealed two sharp fangs.

My chills were short-lived, drowning beneath a wave of curiosity.

Fangs. Actual fangs, just like the stories said. I wondered how that worked? Did his saliva contain an anticoagulant or—

"Would you like to tell me what you're doing in my house?"

He had an accent, a sharp lilt stabbing into the *t*'s and *∂*'s, rising the long *a*'s and *o*'s with a melodic twang.

Interesting. I'd never heard an Obitraen accent before. Then again, most people in the human lands never met anyone from Obitraes, because vampires didn't often leave their homeland and were usually better off avoided if they did.

"I was looking for you," I said.

"So you come into my home uninvited?"

"It would have been easier if you had come to the door."

He paused at the bottom of the stairs. Again, that vampire stillness, the only movement a single slow blink.

"Do you understand where you are?" he asked.

That was a stupid question.

Maybe he was used to being cowered at. I did not cower. Why should I? I'd already met death three times now. So far, the fourth was a bit of a disappointment.

"I brought a gift for you," I said.

His brows lowered slightly. "A gift," he repeated.

"A gift."

He cocked his head, a slow curl brushing his lips. "Is the gift you?"

Another chill up my spine, and this time, I shifted a little to ease it—which I hoped he didn't see.

"No," I said.

"Not this time," he corrected, which I had no idea how to respond to.

"The gift is very special. Unique. You're obviously a man who appreciates unique things." I gestured to the walls and the many artifacts that lined them. "In exchange, I ask you for a favor."

"That isn't a gift," he pointed out. "That's payment, and I offer no services for sale."

"Semantics," I said. "Hear my offer. That's all I'm asking."

He frowned at me, silent. I wondered if someone better at reading faces would be able to tell what he was thinking, but as it was, I certainly couldn't.

After too long, I cleared my throat uncomfortably.

"Is there somewhere we can sit?" I asked.

"Sit?"

"Yes, sit. You must have lots of chairs in here. You must do nothing but sit, being in this mansion all by yourself all day and night."

"Do I look like I do nothing but sit?"

He took another step closer, and I looked him up and down without really intending to.

No, he looked like he did a lot of moving. Probably sometimes lifting heavy things.

I sighed, aggravated. "Fine. We can talk here in the doorway if you want."

He seemed like he was considering it, then acquiesced. "Come."

HE BROUGHT me to a sitting room, which was even more cluttered than the entryway. This one, thankfully, was lit, albeit dimly, with lantern sconces that held peculiar blue flames. Paintings and shields and swords and scrolls plastered the walls. Overflowing bookcases were shoved into every corner—even in front of the windows—and the center of the room was full of mismatched fine furniture. Statues loomed over us—a jade cat staring us down from one side of the room, and a fierce, very naked woman rendered in black marble eyeing us warily from the other. The curtains were cerulean silk, and matching sweeps of fabric hung across the opposite wall, pulled back to reveal another expanse of paintings.

It was a mess, and it was the most breathtakingly beautiful place I had ever seen.

In two seconds, I identified art from four different countries in separate far reaches of the world. The sheer amount of knowledge in this room—I couldn't even imagine.

My eyes must've gone a little wide, because he made a low noise that almost resembled a chuckle.

"You dislike my decorating?"

Dislike?

I considered telling him, *This is the most incredible place I've ever been,* but thought maybe now was not yet the time to start stroking his ego.

"What House are you?" I asked, instead.

Another blink. "Excuse me?" he asked, like he thought he misheard me.

"Which House? From Obitraes." I gestured to the wall.

"This all seems too brightly colored to hail from the House of Shadow. And you seem far too sane to be from the House of Blood. So does that mean you're from the House of Night?"

His brows lowered again, now pressed so low over his amber eyes that they looked like two little jewels peering from pits of shadow.

I didn't even need to question whether that was confusion. Good. Maybe he was surprised that any human cared to know about the three vampire kingdoms of Obitraes. But I liked making it my business to know things. It was the only thing I was any good at, and besides, when you don't have much time in this world, you want to fill it with as much knowledge as possible.

He said, "Are you really not concerned that I'm going to eat you?"

A little, a voice whispered in the back of my head.

"No," I said. "If you were going to do that, you would have done it by now."

"Maybe there were other things I wanted to do first," he said in a tone that implied this often got much more of a reaction.

I sighed wearily.

"Can we talk?" I said. "We don't have much time."

He seemed a little disappointed, but then gestured to the sitting room. I took a seat in a dusty red velvet chair, perching lightly upon it with my back rod-straight, while he settled into the opposite leather couch in a lazy lounge.

"Are you familiar with Adcova?" I asked.

"Familiar enough."

"An illness is plaguing the city."

His mouth quirked. "I had heard that one of your fickle gods had taken a bit of offense to that place. Shame."

As if Nyaxia, the vampires' exiled goddess, was any kinder

of a god than ours. Yes, the twelve gods of the White Pantheon could be cold and fickle, but Nyaxia—the heretic goddess who had split from the Pantheon two thousand years ago to create her civilization of vampires—was just as ruthlessly cruel.

"The illness is getting worse," I said. "It is starting to expand to nearby districts. The death toll is in the thousands and will only rise."

I blinked and saw dust—rancid dust, swept from sickhouse floors and streets and bedrooms. Swept five, six times a day from the church floors, funeral after funeral.

I saw dust that I swept off of Mina's bedroom floor, a little thicker each day. The dust we both pretended did not exist.

I cleared my throat. "All of Adcova's and Baszia's top scientists and doctors are working on finding a cure."

And priests, and magicians, and sorcerers, of course. But I'd given up on thinking that they might save us. It was their god that damned us, after all.

"I think that you, Lord—" I stuttered, realizing for the first time that I had never actually asked for his name.

"Vale," he said smoothly.

"Lord Vale." I clasped my hands before me. "I think that you might have the key to a solution."

He smirked at me. "Are *you* one of the country's 'top scientists and doctors?'"

My jaw tightened. I had always been bad at reading people, but even I could recognize that he was mocking me. "Yes. I am."

Again, that wrinkle between his brows.

"What?" I snapped. "Do you want me to be more demure about it? Are *you*, about your accomplishments?"

Vale didn't look like he was especially demure about anything.

"What is your name?" he said. "In case I need to verify your credentials."

"Lilith."

"Lilith...?"

"Just Lilith. You gave me one name, so that's what I'll give you, too."

He shrugged a little, as if he couldn't argue with that.

"So, Lilith. How do you intend to save the world?"

There it was again—that cloying coating of saccharine mockery, so thick that not even I could miss it.

I said, "I need your blood."

A long silence.

And then he laughed.

The sound was low and restrained, and yet, so thick with unmistakable danger. I wondered how many people had been given that laugh as their final goodbye to this world.

"*You* came here to ask for *my* blood," he said.

Alright, fine. I could see the irony.

"Yes," I said. "I won't need too much. Just a little."

He stared at me incredulously.

"It won't hurt," I said. "I promise."

"I wouldn't think it would." He straightened, crossing one leg over the other.

"I would only need four vials of blood each time. Maybe a little more, if I need extra for additional tests. I would need to come once per month."

He said, without hesitation, "No."

I cursed silently to myself.

"Why not?"

"Because about two centuries ago, I decided that I would never again do anything I didn't want to do. And I don't want to. So no, mouse. That is your answer."

I honestly didn't know how to respond to this. He'd

seemed to be having such a fantastic time toying with me that it hadn't occurred to me that he'd flat out refuse—at least, not so unceremoniously.

His face was a mask now. No wrinkled brow, no smirks. He spoke like he'd just turned down an invitation to dinner from someone he disliked. Pure indifference.

My fingers curled, and I pressed my hands against my skirts to hide the whitening of my knuckles.

Of course none of it mattered to him. What else could I expect from a creature like him—a creature that did not understand life, death, or suffering—but indifference?

I forced myself to do what Mina would do. She would smile sweetly and charm. I was never good at being charming and didn't see much point in it most of the time, but it was worth a try. So I smiled, though it felt more like a baring of teeth.

"You didn't let me complete my offer, Lord Vale. In exchange for your blood, I'll give you a gift each visit."

I reached into my bag and withdrew the rose that I had so carefully packed. I had to stare at it for a moment before I handed it to Vale. Did I imagine that it seemed even more beautiful in here, as if it was meant to exist in this room?

He stared at it, face stone.

"A flower. Very pretty."

He did not even try to hide how unimpressed he was.

"I promise you," I said, "its beauty is by far the least interesting thing about it."

"Oh? And why is that?"

"You won't know unless you accept my deal."

His eyes narrowed at me.

"How many?" he asked.

"Visits?"

"Roses."

"I'll visit you six times, and I'll bring you a rose each time."

This time, I was expecting another unceremonious refusal. But instead, Vale examined the rose, twirling it slightly between his fingertips. He had a very cold, hard stare. It looked a bit familiar, and I couldn't place why until I realized that it was the stare of a scientist, someone used to analyzing things and taking them apart.

A little spark of relief came with this realization. Because that, at least, was something I understood. Maybe Vale and I were worlds apart in every way—human and vampire, lord and peasant, near-immortal and pitifully ephemeral—but if we had that, it was already more than I had in common with most of the people I'd grown up with.

"Fine," he said, at last. "I accept your deal. Did you bring your equipment? Let's get this over with."

OF COURSE I had brought my equipment. I had my needles and vials ready. Vale pulled up his shirt sleeve and extended his arm to me, and I drew his blood.

Up close, he smelled like jasmine—both old and young at once, foreign and familiar. His skin was smooth and tan. When I touched his wrist to adjust the position of his arm, I jumped at the lack of warmth, but it also wasn't as cold as I'd imagined it would be. People spoke of vampires like they were walking corpses, but I'd seen many, many corpses, and Vale didn't look like any of them.

Still, I wasn't quite sure what I was expecting when I pierced the smooth skin of his inner arm with my needle. I had to push much harder than I did with a human, and when the needle went through, it did so with a faint *pop* and abrupt force. The blood that flowed into my vial appeared to be the same consistency as human blood, but much, much darker—nearly black.

I watched it, fascinated. Then, by the second vial, my eyes had drifted up to the rest of the room, taking in the tapestries on the walls, the books on the shelves. Gods, some of those tomes looked to be many centuries old, carelessly shoved into dusty corners.

How old was Vale, I wondered? Legend said he had been here, beyond the outskirts of Adcova, for nearly two hundred years. How many decades—centuries—of life had he lived before then?

How much had he experienced?

"Are you enjoying looking?"

Vale's voice startled me. My eyes flicked back to him. He was now looking at me as he had looked at that rose—pulling me apart, petal by petal.

Are you? I wanted to say.

Instead I said, "What will become of all of this when you die?"

"I'm immortal."

I scoffed. "You're not immortal. You're just very long-lived. That's an important distinction."

"By the time it matters, I'm sure I won't care."

It already looked a bit like Vale didn't care, judging by the condition of his living space, but I didn't say that, either.

A knot of jealousy formed in my stomach. He spoke with such blasé carelessness about all of this. About his life. The gluttony of it revolted me. He'd hoard all of this knowledge here, and he'd think nothing of it. Selfish.

"I imagine it must become the only valuable thing, after all that time," I said. The last vial was almost full. I watched the blood bubble up in the glass, ready to pinch off the needle. "Knowledge."

"Knowledge is cheap and dull," Vale said, too casually, and I almost gasped at him in horror.

"I can't imagine that ever being true. There's so much to learn about the world."

He laughed a little, condescendingly, the way one laughs at a stumbling kitten. I corked the last vial and withdrew the needle from his arm. I found, with some surprise, that his skin had already healed around the needle tip. I had to rip it from his vein, which he didn't react to.

"After so long, you realize that knowing things doesn't especially matter very much. Knowledge with no context is meaningless. That's not the real treasure."

"Oh?" I tucked away my tools and stood. "What is, then?"

Vale stood, too. He was quite tall, and he looked down at me with a wolfish kind of delight. He smiled, revealing those deadly fangs. The moonlight from the window glinted in his amber eyes.

I felt, all at once, like an idiot for thinking before that he didn't look monstrous. Because in this moment, with that smirk on his lips, I glimpsed the man of the legends. The monster of the whispers.

"Curiosity," he said.

PART TWO

THE SECOND ROSE

CHAPTER FOUR

V ale's blood was beautiful. There was no other way to describe it—it was as undeniably aesthetically pleasing as a field of flowers.

It was almost dawn by the time I returned home that night. I wasn't tired, though—no, far from it. I was literally shaking with excitement, my mind running over every moment of that visit over and over and over again, burning it to memory. I hugged my pack to my chest for most of the walk, as if to shield it from the world. It was contraband, after all.

When I got home, I went straight to my office and bolted the door behind me. I didn't need Mina knowing what I was up to, both for her sake and mine. The less I involved her in my blasphemous little scheme, the better.

But there were no footsteps in the house yet. Mina was still fast asleep. I pulled out my instruments, messing up everything I had been so careful to neaten before my departure. I dragged a side table to the center of the floor, setting my seeing lens atop it—a device comprised of many brass rings stacked on top of each other, the top one on hinges and covered in glass so that it could be positioned upright. Runes and sigils had been carved into each ring of metal, and when I touched it, I could feel the magic pulsing from it. I grabbed my ink and stuck my finger into it, drawing a series of marks around the outermost circle of the device.

I didn't have a shred of magic myself, of course, nor did I especially want any—I'd seen many times how it could lead to ruin. But the tools magic could produce were undeniably useful. This one had been created by a priestess of Srana, the Goddess of Seeing and Knowing. I did like to see things, so at least I could be grateful to Srana for that.

I finished the runes, placed my vial at the center of the device, and blew out the candles. The uppermost ring of copper glowed with steady warmth, and when I adjusted the hinge, a ring of light was cast upon the wall.

Within that ring was Vale's blood—his blood at its most base level, the tiniest particles of life within him. They looked like a field of red-black flower petals across the plaster, moving in slow constellations like the stars across the sky.

Sometimes people talked of vampires as if they were living death, nothing more than animated corpses. One look at Vale told me that wasn't true. Still, I knew that vampires had a closer relationship to death than humans did, so perhaps I might have expected to see some of it in the makeup of Vale's body.

No. None of this was death. It was beauty and life and an astounding miracle. He was hundreds of years old and yet his blood was healthy and thriving. It was graceful, elegant. It looked so different from human blood, and I was certain that it would react differently to every test. And yet, there was something so familiar in it too, as if we had been the originals and he had been the improvement.

Maybe the vampires' heretic goddess had been onto something, after all.

I stared for far too long, transfixed.

My instrument had been created with the magic of Srana, a goddess of the White Pantheon—the White Pantheon that despised Nyaxia, the mother of vampires, which meant I had to be very careful with the instruments I used around this blood.

Even the fact that I had it at all... here, in a town that worshipped Vitarus...

I blinked and saw my father kneeling in that field of death, knuckles trembling around a fistful of doom, ready to spite a god that would happily spite him back.

I pushed the thought away and quickly broke down the instrument, tucking Vale's blood into a drawer.

Still, I couldn't help but take it out every few hours to peer at it, even if only for seconds at a time. I told myself it was for work—and it mostly was, because I didn't stop working for more than ten minutes at a time those next few days—but really, I was... well, a little transfixed by it. Every time those splotches of black lit up my wall, I released an exhale of awe.

"What's that?"

I spun around. Mina stood in the doorway. For a moment, in contrast to the elegant vitality of Vale's blood, the sheer withering mortality of her shocked me. Darkness ringed her eyes and dusted the deepening hollows of her cheeks. Once, she had been a strikingly beautiful girl—and she still was, but now hauntingly so, like the face of a stone goddess at a grave site. I glanced down. How long had she been here? I wasn't sure which answer was worse. Longer, and she saw more of what I was doing. Less, and I could be more concerned about the distinct layer of dusted skin that already coated the floor around her feet.

"What's that?" she asked, again.

"Nothing," I said, even though my sister knew me well enough to know when nothing meant everything.

I thrust the vials and my lens into my bag, buttoned it, and rose.

"I have to go," I said. "I'm visiting Farrow. Rosa will be by with dinner for you, and —"

I stepped past her, but Mina barely moved aside. When I brushed by her, I tried not to notice the faint fall of fine dust to the floor, steady as seconds ticking by.

"Lilith, wait —" she said.

I stopped, but did not turn back.

"What?"

I sounded colder than I wished I did. I wished I could be warm like Mina was. Like our mother had been. Our father. In a family of warmth, I was the strange, cold one — the one who could decipher textbooks and equations but struggled to decipher the exact cadence of a voice that made a name a term of endearment, nor the pattern of a touch that made it a caress.

"Stay with me today," she said. "We can take a walk."

"I wish I could. But I have too much work to do."

Even I knew how to recognize the frustration in her voice when she said, "Why?"

I knew what she meant: *What could be more important?*

Growing up, people would always ask me, *Why do you work so hard?* They would always ask in the same tone of voice — confused, pitying — the kind of tone that told me they were asking me a different question than their words alone conveyed. In that tone, I heard all the implications. The implication that I was wasting my life. I had so little of it, after all. Why spend it toiling away?

I heard that in Mina's voice now. That same judgment,

same confusion. Except now she was the one whose time was running out, begging me to take some of it from her.

And that, in the end, was the answer.

Why was I working so hard? I was working so hard because none of it would ever be enough. I would continue until I had nothing left to give. Force myself through the grinding machinery of the mind.

Better this than to spend time making it harder for her to say goodbye to me one day. My love gave my sister nothing. But my work gave her a chance.

"I have to go," I said again, and left Mina in the hall, watching after me.

She wouldn't understand if I tried to explain it to her. She didn't know death like I did. After all, she was never the sister who was supposed to die.

CHAPTER FIVE

I walked to the outskirts of town, where I could catch a boat out to Baszia. On my way through the city streets, I passed a congregation of Vitarus's acolytes kneeling in the streets, praying over piles of burning leaves. At their front was Thomassen, Adcova's head priest and devoted follower of Vitarus—a tall, thin man in his mid-fifties with kind eyes. He was the same age my parents had been, but looked much older these last few years. It must take a toll, spending all this time trying to understand why your god turned against you.

He gave me a faint smile as I passed, which I returned with a curt nod.

He had been good friends with my father, once, so he had always been kind to Mina and I. Pitied us, maybe. Funny, because I certainly pitied him, kneeling in the ashes he fed his god, while his god only gave him more ashes in return.

I continued to the outskirts of town, where I found a ship headed across the channel to the city. The journey took hours, and my stomach had become unaccustomed to sea travel, but when I stepped foot on the docks it was all worth it.

I inhaled a great deep breath of the city air—air that seemed to smell of books and excitement and knowledge... mixed, maybe, with just a hint of piss. I'd spent six years here, studying at universities and libraries. Right now, it struck me with staggering force just how much I had missed it. Even the buildings, tall and majestic, spoke of history—many of them had been erected a thousand years ago.

Farrow was, as I knew he would be—as he always was—in his study, a little room tucked in the back of the university archives. And, as I knew he would be, he was happy to see me. The brightness of his grin when he looked up to see me lit a little spark of guilt in my chest.

I shouldn't be doing this. Shouldn't be putting him in danger.

But he was one of the most intelligent people I knew, and I needed help.

Farrow was tall and slender, with ash-blond hair that he constantly had to push out of his eyes and silver glasses that were always a little broken. He had a way of bending his whole body up with interest in whatever he was working on, and that was exactly what he did as I set up my lens in the center of the room, folded over at the very edge of his chair as I projected those beautiful blood flower petals onto his chalk-smeared wall.

His eyes widened, and he crept a little out of his chair to look closer. He barely even breathed.

I did always so appreciate that about Farrow—his unabashed amazement at the world around him. When I first met him, as a young student, I had loved that he embodied what I myself so wanted to express but couldn't. Men from upper-class families were welcome to be openly delighted by their craft. It made them interesting and eccentric, committed and passionate. When women did it, it made us vapid.

I had seen Farrow amazed many, many times. But never so much as he was now. He rose, circled the room, squinted at the blood, then eventually returned to his chair and sagged into it, running his hands through his hair as he peered at me from behind askew glasses.

"Great gods, Lilith, what *is* this? What am I looking at?"

I swallowed. I didn't want to say the word, not aloud. It would only put Farrow at risk—more than I already had by coming here. There was a reason I had brought it here, rather than asking him to come visit our cursed town.

Selfish of me. I knew that.

"If I could distill this somehow," I said, "how would I do that? This property?"

"You would need magic, probably."

"What if I couldn't do that?"

Farrow frowned. "Why would you not be able to do that?"

I eyed my lens. The projection had been up for longer than I'd ever allowed myself to look at the samples at home. I feared that at any moment, the magic would recognize the nature of what it analyzed. Magic was fickle and temperamental, just like the gods.

"Could there be a way to do it without?" I asked. "By scientific rather than magical means?"

Farrow seemed confused, which was reasonable. Science and magic were often two parts of the same whole—each complemented the other, their methods often inextricable.

"It would be... it would be hard. Maybe impossible. Bring it to one of Srana's temples. See what the priestesses have to say about it."

"I can't."

A wrinkle formed between his brows. His amazement had faded. "*Why*, Lilith?"

My teeth ground. I swept the runes from the table in one

abrupt movement just as the lens began to smell faintly of smoke. The room went dark as the projection flickered away. Even in the dark, I could feel Farrow's stare, hard and piercing. Gone was his childlike amazement. Now he seemed only concerned.

Of course he was. I had come here because he was one of the smartest men I knew. Could I really expect that he wouldn't figure out what was sitting right in front of him?

"Thank you, Farrow," I said. "I appreciate your—"

I turned to the door, but he caught my arm, gripping it tight.

I looked down at his long fingers around my forearm. Then up at his face. His concern now overtook his expression with the same enthusiastic verve that his joy had minutes ago.

That was Farrow, of course. Feeling everything. Showing everything.

"Tell me what this is," he said.

I shook my head, and that was answer enough for him.

"It's one of them, isn't it? Alive and dead at once. I thought it at first but then I thought—you couldn't have—" He swallowed, his fingers tightening. "Tell me I'm wrong."

I couldn't. I had always been such a bad liar.

His expression sank with realization. He went pale. "What are you doing, Lilith?"

I pulled away, his judgment burying deep in my gut. "I'm doing what I have to."

But with Farrow, once questions started, they never stopped. "How did you get this? How—" Another wave of realization. "*Him?* You went to visit *him? By yourself?*"

"I did what I had to do," I hissed, again. I struggled to hide my annoyance.

Did he think I didn't know that it was stupid? That it was dangerous?

He whispered, "You want to *inject people with vampire blood?*"

I spun around. "*Sh.*"

His mouth snapped closed. Our eyes both flicked to the ceiling—to the sky beyond. The gods, after all, were always listening.

I drew in a deep breath and let it out.

"It won't be that anymore by the time I'm done with it," I said quietly. "I'll make it into something different. Into medicine."

He shook his head sadly.

"This is dangerous."

"I will make sure it doesn't hurt them."

"Them? What about *you?*"

I couldn't dampen my frustration. "So what?" I snapped. "We've prayed. We've used the arts of the White Pantheon. We've given Vitarus every offering for ten gods-damned years. We have listened to the scholars and the priests and the acolytes and the sorcerers. We have tried the magic of every god, including the one that damned us. What the hell else can we do?"

"I'm worried about you, Lilith," he said, softly, and I wanted to laugh at him at first, because what good was worrying about me and my pitiful remaining months of life compared to the fate of an entire city?

He drew a little closer, close enough that I could feel the warmth of his skin, our faces now just a couple of inches apart.

"I'm worried about you," he said again, more softly—more tenderly. "Stay. We'll figure something else out together."

Stay. It wasn't the first time he had asked me that, the first time he'd offered me that word with his breath close enough to warm my mouth. *Stay.*

And just like the last time he'd said it, I was tempted.

There was a certain comfort in Farrow. I liked him. I trusted him. I knew how he kissed, how he touched, how he felt within me. I spent so much time thinking about every muscle in my body, my face, figuring out how to present them appropriately to the world. Despite all that effort, most people still didn't like me much, but Farrow always had. And when he told me, *We can figure it out together,* I didn't believe him, but he made me want to. And that counted for something, didn't it?

But it was unfair to Farrow to let him love me the way he wanted to, or to let myself love him with the fractured pieces I could offer him. He always wanted more than I could or wanted to give him. It is, after all, a waste to love a thing that will soon be gone.

I pulled away from him and picked up my bag. Farrow looked at it, and I knew he saw how well packed it was — packed enough for a much longer journey than the one back home.

"Are you going there now?" he asked, alarmed.

"Yes."

"Lilith." He sounded hopeless, broken. I didn't want to turn back, but I did anyway.

"Stay," he said again. "Please."

"What are you afraid of?" I said. "That I'll fall into the darkness? *We're already there.* We've been there for years, and we're only falling deeper."

I did not add: *And I've been there since the day I was born.*

I shook my head. "It's too late, Farrow. It's too late."

CHAPTER SIX

It was long past midnight by the time I reached Vale's mansion. It was drizzly and cold, as it often was this time of year. I knocked on the door and received no answer.

I was tired and damp, uncomfortable and oddly on-edge after my encounter with Farrow. I was in no mood for games.

I pounded hard on the door, five six seven eight *nine times*, and when there was *still* no answer, I opened the door myself. Vale still didn't lock his door. Why would he?

"Lord Vale?" I called out into the cavernous darkness as I closed the door behind me. I heard nothing, saw no movement. Perhaps Vale had decided he was tired of me, and he'd ignore me until I went away. Or maybe he'd lure me in and wait until he could grab me and devour me.

I wandered through the master hallway, and when I found nothing, decided, after a moment of hesitation, to climb the stairs.

I told myself that I was simply accomplishing a task—but

if I was honest with myself, I'd acknowledge the little trill of delight that ran up my spine.

My mother used to say that I enjoyed the sciences because I was a naturally nosy person. She was probably right. She had always known me better than anyone.

I collected facts the way other people collected jewels, and Vale's home was overflowing with them—both facts *and* jewels. The stairs led to a long hallway, just as cluttered and architecturally dissonant as every other part of the house that I'd seen so far. The walls were lined with artwork, most of it depicting vampires with feathered wings gutting, stabbing, burning, and otherwise brutally killing their victims—most often vampires with bat-like wings. But these halls also held other artifacts, too. One stretch displayed a set of grand wing bones, which unfolded along the peeling gilded wallpaper. I had to pause to stare at them in awe.

Incredible.

I'd never seen such a work of biological art. Each wing was longer than I was tall, the bones a delicate gleaming ivory. But despite their light elegance, they were also clearly powerful—even without muscle, I could see that.

I must have been right about Vale. He must be a Nightborn vampire from the House of Night—the kingdom of the only winged vampires.

What did *his* wings look like?

A distant voice jerked me from my thoughts. I tensed, face snapping to the end of the hall.

The sound had come from around the corner, and it came again. A voice, I realized after a moment—though too high to be Vale's, and wordless. A cry. Pain?

My heart quickened a beat.

I hadn't thought much about whether Vale did indeed eat

humans. And if, when he did so, he dragged them back here to do it.

I probably should have run. But there was no use fighting nature, and I was a curious creature. So I went not away from the sound but closer, creeping down the hall and around the corner, where cool lantern light spilled from an open door at the end of the corridor.

The sounds grew louder, closer.

And a flush rose to my face when, a few steps away from the door, I realized that what I was hearing were not cries of pain. Very much the opposite, actually.

The moans rose to a crescendo.

No, Vale was not alone. And whoever he was with was having a wonderful time.

The door was wide open. Who could blame me for looking?

I peered around the frame. It was Vale's bedchamber, a grand room covered in silks and art, with messy trinkets strewn over each surface. A large bed with a carved frame sat in the center of the room. Fine bedsheets were mussed and tangled over it.

And tangled over it, too, were two figures so entwined I wasn't sure where one of them ended and the other began.

She was beneath him, an expanse of golden skin gleaming beneath the messy curls of red hair, and he leaned over her and clutched her hips from behind. I mostly saw his back and her tangles of hair, her arms splayed and gripping the bedsheets to brace herself as he drove into her viciously. With every thrust, his muscles flexed beneath his skin, rippling over the broad expanse of his back, the curve of his backside, the lean muscle of his upper thighs.

He looked as majestic and beautiful as those wings had. I

imagined that perhaps, covered in muscle and skin, they might look almost—*almost*—as beautiful as he did now.

My face was very hot.

I couldn't look away. I really did mean to announce myself, or back away, but I found myself frozen.

The woman bent down against the bed, the pillow slightly —but only slightly—muffling her rising cries of pleasure. Vale's movements grew faster, harder, flesh slapping against flesh, leaning against her and falling over her back.

I watched, unblinking, as he held her down, mouth going to her shoulder as they came together. He made a sound only then, a rough exhale that made the hairs rise on my arms, and I had to strain hard to hear it over the sound of her.

They collapsed together, and with their breath, I let out my own. My fingers loosened around the doorframe. I hadn't realized I'd been clutching it.

Vale whipped around.

"Lilith."

For just a split second, he actually looked shocked. Frazzled.

Then his face hardened, going smooth and angry. He turned his back to me and rose from the bed, yanking a crumbled-up pile of fabric from the floor and giving me another distracting view of his backside.

"What," he snapped, "are you doing here?"

"You didn't answer the door."

My voice sounded a little weaker than I would have preferred.

The woman made no attempt to cover herself. She rolled over and stretched. I realized that she was covered in blood, especially around her throat—the dark color of the bedsheets had hidden that from me before. She smiled, revealing pointed teeth.

"You invited a human friend, Vale?" she said, with a deep inhale that had me stepping backwards.

Vale shot her a warning glance that made her smile disappear.

"A mouse," he sneered. "No, a rat. An uninvited pest."

He shook out the robe he'd picked up with a single violent movement, then threw it over his shoulders.

"I knocked," I said. "You didn't answer. I came when I said I would."

"Oh, so did I," the woman said, laughing softly to herself, and Vale shot her another unamused stare.

"What?" she said. "You don't want to share?"

"Let's not make any more a mess of my home than we already have. Can you give us a moment?"

She sighed, then sprang from the bed, lithe as a cat. She grabbed a piece of fabric from the bedside table and wiped the blood from her chest and throat. "I should be going, anyway. Thank you for the hospitality, as always, Vale."

She threw on a plain black shirt and trousers, which had been sitting on the ground, then strolled past me with nothing more than another long, curious stare, which started at my feet and ended at my face.

Vale stared out the window, silent, until her footsteps had long since disappeared. Then, finally, he turned. He now wore a dark red, velvety robe, which he had loosely tied around his waist, so it revealed a long strip of his chest—covered in curly black hair—but, almost disappointingly, nothing below his waist.

My lips pressed together.

The robe was so...

"What?" he snapped.

"What?"

"You're laughing at me."

"I'm not laughing at you. I'm laughing at — "

I closed my mouth. Telling people that I was laughing at their clothing, I realized, was probably not very polite.

"*What?*" he bit out, irritated.

"The robe. It's just... it's very vampiric."

His lips went thin. "Yes, well. I am a vampire. So I see now why you're at the top of your field."

I stifled my laughter.

Right. Work.

"I'm here for your blood. It's been a month, as we agreed."

"And payment?"

I reached into my bag and withdrew a rose, carefully wrapped so not a single petal was bent or crushed. He outstretched his hand, and I hesitated, to which he heaved an irritated sigh.

"What? *Now* I scare you?"

He didn't scare me. It just smelled like sex in here. I crossed the room, eyeing the bloody, rumpled sheets as I passed. Vale took the rose and stared at it, unimpressed.

"The one you gave me last time seems to be totally unremarkable," he said.

"You'll have to be patient."

"I'm not a very patient man."

"I don't lie, Lord Vale. They're special. I promise."

"You can just call me Vale," he grumbled. "I suppose that once someone has seen my bare ass, we can drop the titles."

He dropped heavily into a velvet chair next to the window. "Let's get this over with."

"Here?"

"Yes, here. Is that a problem?"

I glanced again to the bed, and he let out a low, silken chuckle.

"What? Are you really so distracted by sex?"

41

It *was* distracting, but I wasn't about to admit that. I dropped to my knees before him and withdrew my equipment from my bag. When I took his arm to guide the needle into his veins, I was acutely conscious of every patch of my flesh that touched his.

He laughed again as I thrust the needle through the resistance of his skin.

"I can hear your heartbeat. Is that nervousness or excitement?"

I could hear my own heartbeat, too, and I wished it would calm down. Even I wasn't sure which it was, but neither was welcome.

"I think it's amusing that you wandered into my house without a care in the world," he said, "but the sight of fifteen seconds of sex triggers your nerves. I will never understand humans."

"I've had plenty of sex." And the minute I said it, I cursed myself for it—why in the gods names did I just say that?

Vale now looked very, very amused, and I absolutely despised it.

"Have you, now? Did some gawky farm boy from next door take you for a ride?"

My lips thinned.

Eron had been gawky, and he was a farm boy, and that summer when I had been sixteen and curious, we had indeed explored each other in the deserted moments behind the barn, when no one else was around. I didn't want to die a virgin. I was certain, then, that I wouldn't live to see the winter, so I saw all of Eron instead.

But fifteen years later, I was still here, and six months ago, I swept Eron off the church floor after his funeral, when his mother was too hysterical to do it.

"You know, I did wonder at first," I said, "why you didn't

42

kill me when I came into your house. Now I understand it's because you're a bored, lonely man, desperate for any kind of company."

I didn't look away from the vial, his blood dripping and rolling against the glass. But I felt his stare, and in the moment of silence, I wondered if I'd hit my mark.

"As you just witnessed," he said, coolly, "I can get all the company I want."

"Company that got what she wanted from you and then left without saying goodbye."

"We got what we wanted from each other. It wasn't conversation that I was looking for."

And yet... he was sitting here talking to me.

"What do you need this for?" he asked. "The blood?"

"As I told you—"

"My blood isn't a cure for anything, I promise you that."

"It appears, L—" I caught myself. "Vale, to be a cure for death."

He scoffed. "No human encounter with vampire blood has ended particularly well."

That tone piqued my curiosity almost enough to make me forget my irritation at his insults. I peered up at him. He was looking out the window now, the cold moonlight tracing the outline of his jawbone, especially strong from this angle.

"Were you Born or Turned?" I asked.

There were two ways to make a vampire. Some were birthed, just like the rest of us. But more interesting was Turning—the process of drinking a human's blood, and offering theirs, to create a new vampire.

I'd thought a lot about it these last few weeks. What that must be like. What other animal could do that? It was a transformation as impressive as a caterpillar becoming a butterfly.

His gaze shot to me, insulted. "Born. Obviously."

"Why is that obvious?"

"Being Turned is... undesirable."

I knew only a little about vampire anatomy. It was difficult to study them when they were so reclusive. And when so many of the humans who went to Obitraes never returned.

"Turning is dangerous, isn't it?" I asked.

"Yes. The majority die during the process."

"But if someone survives it and becomes a vampire, they're considered... undesirable?"

"Part human. Part vampire. Their blood will always hold the taint of humanity." His nose wrinkled. "Less pure."

"But if they survived such a dangerous thing, doesn't that make them the strongest among you?"

Vale opened his mouth as if to argue with this, then shut it. He looked conflicted, like he'd never thought of it that way.

"It's just not how it is," he said, at last.

The first vial was full. I switched to the next.

"Why did you leave Obitraes?" I asked.

"And I thought you were nosy last time."

"Most humans never get to speak to a vampire. I should take advantage of it, shouldn't I?"

"Aren't you so very lucky."

A few seconds passed. I thought he didn't want to answer, but then he said, "I wanted a change."

"Why?"

"Why not? Have you always lived in that little town?"

"I studied in Baszia."

He scoffed. "A whole ten miles away from home. How *exotic.*"

I did despise that he was so judgmental, and I despised even more that his sneers prodded at a selfish little wound I tried to ignore. I would never get to see the world—but that didn't mean I didn't want to.

"Not all of us have the resources to travel," I said.

"Humans and your money."

"I didn't say money. I said resources."

He glanced at me in confusion. I gave him a grim smile.

"Time, Vale," I said. "Time is the most valuable resource of all, and some of us are perpetually short."

CHAPTER SEVEN

Vale led me back downstairs when I finished collecting the blood. As I did every time, I found myself slowing down every hall, unable to look away from each antique and piece of art. I couldn't stop myself from craning my neck as we passed the wings again, my steps slowing without my permission.

"You like them?"

Vale sounded amused.

"They're... remarkable."

"More remarkable on my wall than they were on the man who bore them."

It was a horrible thing to say. A reminder of vampire brutality. And yet... I was intrigued more than appalled.

"And who was that, exactly?"

"A Hiaj general who was said to be *very talented*."

The words *very talented* dripped with sarcasm.

"Hiaj," I repeated. "That's one of the two clans of the House of Night?"

My gaze traveled to the painting beside the wings—

depicting a man with feathered white wings driving a spear through the chest of another with slate-grey bat wings.

"You... know more of Obitraes than I'd expect of a human."

"I like knowing things."

"I can see that. Yes. Hiaj." He tapped his finger to the bat-winged man. "And Rishan." He tapped the feather-winged man.

Rishan. I formed the word silently, rolling my tongue over it.

"You must be Rishan, I assume. Going by your taste in decor."

"You assume right."

"So you have wings."

I said it before I could stop myself. Feathered wings. What would they look like? Would they be dark, like his hair?

"You're an especially nosy mouse today."

I blinked to see Vale staring at me, amused.

"I'm always nosy," I said. "You don't know me very well yet."

Yet. As if we would form some kind of friendship through this little bargain of mine. A ridiculous thought. Still... when he laughed a little and grinned—reluctantly, like he didn't mean to—I could imagine it could happen.

"Maybe you'll get to see them one day," he said, "if you're very fortunate."

And I could imagine, too, that I would indeed be very fortunate if I got to see Vale's wings.

"Who is in power now?" I said. "Back home?"

"Home?"

He said the word slowly, like it was foreign.

It didn't occur to me that Vale might not think of the House of Night as his home. But then again, would one consider a place their home when they hadn't been there for hundreds of years?

"The House of Night," I said. "The Rishan and the Hiaj are always fighting, aren't they? Struggling for power."

"You know too much of my country's dirty laundry."

"I had a colleague once who studied anthropology, with a specialization in vampire culture."

Vale laughed. "A dangerous field."

Dangerous, indeed. He had gone to Obitraes and never came back. He was a nice man. I liked to think that perhaps someone Turned him and he was still living some life over there—even though I knew it was more likely that he just became somebody's meal.

Vale turned and started walking back down the hall, and I'd given up on getting an answer to my question when he finally said, "The Hiaj. The Hiaj have been in power for two hundred years."

So Vale's people had been usurped. Judging by the style of art and what I knew of vampire conflict, that couldn't have been pleasant.

And...

"How long have you been here?" I asked, carefully.

Vale chuckled at the question I really asked and gave me the answer I was really looking for.

"It's not a pleasant thing to oversee the loss of a war, mouse," he said. "You'd move halfway around the world after that, too."

HE INDULGED MORE of my curiosity on our walk back to the front door, pausing here and there to tell me a few facts about this artifact, that painting, this tapestry. Even those sparse tidbits were more than enough to confirm that I'd been right —that Vale had incredible amounts of knowledge just holed up in this castle, never mind what he must own back in the House of Night. But what struck me even more is that he offered this information to me freely, without me even having to ask, like he understood each question I had before I asked it. I would have almost thought he was a mind reader, but the House of Night did not have mind magic—that was a gift reserved for the House of Shadow.

No, he was just... observant. And for those few moments, strangely enough, I felt like I didn't have to work so hard to bridge the gap between myself and the rest of the world. Didn't have to work so hard adjusting my facial muscles and body language, nor at decoding his.

And maybe... maybe he felt the same way. Maybe—for all that my prodding earlier had simply been cruel words to throw at him—maybe he really *was* lonely.

This thought struck me all at once when he went to the door, opened it, and then stopped.

I was so bad at reading expressions. But was that... disappointment?

He stared out into the dark path ahead.

"It's late," he said. "How long does it take you to get back to your home from here?"

"A few hours."

That was an understatement, actually.

"Isn't it dangerous for a little human mouse to travel so far alone at night?"

"It won't be night for much longer."

My body refused to let me forget it, too. Every blink was

gritty, and my muscles grumbled in irritation. I was thirty. Old enough for my body to protest a night absent of sleep in ways it hadn't ten years ago.

But I shrugged.

"If I didn't do dangerous things," I added, "I would do nothing."

"Hm." He eyed the trail, then looked back to the stairs, seemingly unconvinced.

I cleared my throat and adjusted my bag over my shoulder. "Well—"

"You could stay," he said. "If you would prefer to wait until morning to leave."

He looked as surprised to have said it as I was to have heard it.

I arched an eyebrow. "Well, Vale, you already had one houseguest this—"

"Not like that," he huffed. "The houseguest is gone. I offer you your own bed. Though"—and here his voice lowered, slightly—"if you wanted to share mine instead, I wouldn't object to that, either."

I stilled. Words evaded me. I searched his face for any one of the many signs I'd memorized that someone was making fun of me, telling me something that wasn't true, and I found none of them in Vale's expression.

That surprised me almost as much as it surprised me that I was considering it.

That I found myself, far too vividly, imagining what it might have been like to be in that woman's place—to feel his hands over my body, pinning me. To feel the size of him inside me, feel what it would be like to be taken that roughly, that hard. I'd been fooling myself if I thought I had put him out of my mind. If there was any part of me that wasn't thinking, just a little bit, about the sheen of sweat over his bare muscles with every movement he made tonight.

I cocked my head and stared at him.

"Vampires have a good sense of smell, don't you?" I said.

He had moved a little closer. "Yes."

"Do you smell me?"

My voice was low, rough.

"Yes," he said. "Acutely."

"Is it… difficult for you?"

"What does that mean?"

I didn't answer, and the corner of his mouth lifted. "Are you asking if I'm tempted by you?"

He leaned closer still. My back pressed to the doorframe. I remained very, very still, even as he stepped closer, our bodies almost—but not quite—touching. He lowered his head, so his lips nearly came to my throat.

I didn't move.

My breath had gotten shallow, my heartbeat faster. Some primal thing within me reached for the surface of my flesh—reached for the surface of his.

His mouth did not touch me. But I still felt the vibration of his voice, deep and low, over the fragile skin of my throat.

"I smell you," he murmured. "I smell your blood."

"What does it smell like?"

It sounded like someone else's voice.

"It smells like honey. Like… nightshade. Sweet. Perhaps with a bitter bite."

I heard his voice dip a little at that last part. Amusement.

"And?" I said.

"And I smell the beat of your blood through your veins."

My pulse quickened a little, as if stirring beneath his awareness. His hands braced against the doorframe now, so his body enveloped mine—though, still, without touching.

"And you know what else I smell?" His face ducked a little closer, voice lowering to a whisper. "I smell that you want this."

I let out a rough breath.

I did. My curiosity extended beyond artifacts on a wall. It reached for Vale's body and my own, and what it would feel like to bring them together.

I wouldn't even try to deny that to myself.

But I wasn't about to let him take me to bed in sheets still mussed from someone else's body.

"Wanting something doesn't count for anything," I said, and put my hand firmly on his chest, pushing him back. He stepped away without protest, eyes narrowed—maybe with curiosity just as potent as mine.

"Goodnight, Vale," I said. "Thank you for the blood. I'll see you in a month."

And I didn't look back once as I set off down the trail.

I knew he watched me until I was gone, though.

WHEN I GOT HOME, the house was still dark and quiet, though the birds were stirring by then. I called for Mina and heard no answer.

Maybe she left early, I thought, not believing myself.

I found her in her bedroom, perched at the edge of the bed, her eyes glassy and glazed over, her joints locked and muscles tight. She didn't see me even when I stood right in front of her—not until I shook her, hard, and she blinked and finally looked up at me.

"Oh. You're home!"

She hid her fear beneath her smile and a dismissive wave, and even though a knot formed in my throat that made it hard to speak, I didn't question her.

But I still saw her trembling. Still saw the way she paused in the mirror when she rose, shakily, from the bed, looking at herself the way I had the first day I was old enough to feel death following me.

So much of her skin covered the floor that it took me half an hour to sweep it all away.

PART THREE

THE THIRD ROSE

CHAPTER EIGHT

Three weeks of relentless work passed.

I threw everything that I had into it. I stopped sleeping, save for brief naps taken out of sheer exhaustion, and only when my body threatened to betray me. I stopped eating, save for hurried bites of whatever was easiest to shove into my mouth over my books. I stopped leaving the study, save to go cultivate my roses, making sure they remained perfect enough to pass Vale's exacting standards.

"Why are you working so hard?" Mina would ask me sadly, with lips tinted black from the answer to her own question.

I couldn't waste time. Time was precious.

My own condition deteriorated, too, old symptoms that I'd grown used to now creeping up on me with renewed verve. But those were nothing compared to those that nibbled away at my sister's life, bit by bit.

When I closed my eyes, I saw Vale's blood. I stared at it twelve, fifteen, eighteen hours a day, always in small bursts to avoid rejection from the magic of my instruments. It happened anyway, eventually, the glass cracking with bursts of acrid smoke. I had to run into the city to buy another lens for far too much money that I did not have. Not that I cared—who could care about money in times like this?

I began distilling Vale's blood into potions. My early attempts were clumsy, one even erupting into eerie white flames. But after countless trials, my concoctions were no longer smoking or giving off rancid, rotting smells. Eventually, they started to resemble something like actual medicine.

One day, I produced something that responded well to all my tests. It didn't combust, or smoke, or burn. It didn't harm plants or skin. It had all the markers of a potential candidate —and it didn't even resemble blood anymore.

Finally, after much internal debate, I gave it to one of my ailing test rats.

Animals didn't respond to the plague the same way humans did, which made it difficult to test medicine on them. This rat was ill—it had days left, if not less—but it wouldn't wither to dust the same way humans affected by the plague did.

Still… information was information.

I watched that poor rat day and night. Hours passed, then two days. I half expected the creature to die a slow, miserable death.

It didn't happen.

In fact, the rat didn't die at all. Not even when the illness should have stolen its final breaths.

No, it was still lethargic and slow, still obviously unwell, but it did not die.

It was such a tiny, tiny victory—not even a true positive outcome, but the absence of a negative one. Still, that was enough to have me grinning giddily all day. I felt, deep in my bones, that I was getting closer.

I gave up on even trying to sleep that night. It was midnight and very stormy, violent drafts through my office window blowing my candles out every few minutes. But I had work to do.

After only an hour, though, I reached into my pack to find that, in my exhaustion, I'd miscounted—I was out of blood.

I cursed.

I stared at the empty vials over my desk. Then at my dozens of failed experiments and the single—almost—successful one.

I looked to the window, and the ferocious night beyond the glass.

It wasn't even a decision, really.

I rose, gathered my things, and walked down the hall. I peered into Mina's room on my way out. Her sleep was restless, and she left dusty marks on the bedspread.

The sight was far more frightening than that of the storm outside.

Vale wasn't expecting me yet. It hadn't yet been a month. Maybe he'd turn me away. But I couldn't afford to wait.

I tucked a rose into my pack and went out into the night.

IT WAS dangerous to travel in this weather. Rationally, I knew this, but it didn't feel like much of a danger until I was actually stumbling through the soaked, pitch-black forest paths. I spent so much time thinking about death at the hands of my illness that it had become easy to forget that there were countless other ways it could take me, and a night like this was full of them.

It took me twice as long that night to make it half as far. I had to focus absolutely on the road in front of me, trying not to slip on soaked rocks or sink too deep in the muddy dirt. The rain let up a little bit, eventually, but I was so exhausted by then that I wasn't alert.

I didn't see the men surrounding me until it was too late.

One minute, I was dragging myself along the road, and the next, pain burst through my back as a force slammed me against a tree.

Crack! The back of my head smacked wood.

Everything went dull and fuzzy for a moment—even though I refused to acknowledge it, I had already been on the precipice of losing consciousness from sheer exhaustion. That one hit was nearly enough to push me over the edge of it.

I clawed back to awareness, blinking through the haze at the men around me. A young man held me to the tree, hands to my shoulders. Behind him, several others circled like prowling wolves.

One look at them and I knew they were starving. So many people were, these days.

The boy holding me was tall and broad, but he was barely more than a child. It was hard to read his age because of the gaunt angles of his face. Sixteen, eighteen at most.

His expression changed a little when I met his eyes, quickly averting them. Behind him, one of the men approached. Older, bearded. A hard, angry face.

Five of them. One of me. I'd never thrown a punch or wielded a weapon in my life.

I didn't need to be a renowned mathematician to solve that equation. I didn't try to fight back.

"I don't have anything of value," I said.

"Bullshit," the older man scoffed. Then, to the others, "Take her bag."

My heart dropped.

I'd been in such a rush to leave that I hadn't been picky about what I took with me. I had just thrown everything into my pack. My instruments. Useless to these men—they wouldn't even know where to sell them—but everything to me.

"There's nothing you can eat or sell in there," I said.

But they snatched the bag away anyway, rummaging through it. I cringed at the sound of carelessly clinking glass, punctuated by a few cracking shatters.

My heartbeat throbbed in my ears.

"Please," I said. "Please. It's worthless to you. I'll give you—"

Gods, what could I offer them? I had nothing of value to give them in exchange. I had no money on me. Little at home, either. I didn't even think to pack food, not that I thought these men would be satisfied with a single woman's scraps of bread.

The boy, the one who held the knife to my throat, winced again. Guilt? Was that guilt? I so wished I was better at reading people.

"Keep that knife to her fucking throat, Filip," the man snapped, then smiled at me—a horrible expression, like a snarling wolf. "What? What will you give us instead?"

"I—"

My mind wouldn't work. The gears were sticky and slow with exhaustion. He reached for the bag again, and I said, "No. Please. I'll give you double what it's worth once I'm home."

"Once you're home?" the man scoffed. "Oh, I trust you."

The other men laughed. Filip looked pained. My gaze flicked to his, though he avoided looking at me.

Mina would try to connect with him. She'd know what to say to make him let me go.

"Filip?"

His eyes reluctantly lifted to mine.

I should have had some moving plea, some emotional words for him. But emotions and sentimentality had never been my strong suit. Instead, I told him the truth.

"I'm not lying to you," I said. "I will double what that bag is worth. I promise you."

And I did, I really did, mean it.

But the older man's smile curdled to a sneer. "Do you think we're stupid, girl?"

I bit back a surge of frustration.

Why were humans so illogical? I was offering them a good deal. A good trade. More money. And yet, I couldn't make them believe me.

"We'll take your dress instead," the man said.

Filip's grip on the knife loosened again. His head whipped to the man, like he was going to say something and then stopped himself.

I was confused. I looked down at myself. My dress might have been worth something a decade ago. Now it was old and stained, the hem tattered from my journey.

"The dress is worth nothing," I said, annoyed. "I'm offering you a better deal."

"I'll take something I can have now over your empty promises."

"But it's—"

The man snatched the knife away from Filip, thrusting it against my throat. A shock of pain that seemed distant slithered over my skin. Something warm and wet ran down my throat. "I don't need your fucking arguments," the man hissed. "Take it off or I cut it off you."

I was grateful for my irritation, because it dimmed my fear.

"I can't take it off if you don't give me room," I said, attempting to move my hands to my buttons to demonstrate — he was in the way.

The man stepped back reluctantly, pulling Filip along with him.

I looked at the newly opened space between us, a pang of desperate longing in my chest. There it was. Four feet of space between me and my assailants, and endless possibility I couldn't seize.

I had always been quite comfortable with who I was. I was never the athlete, the warrior, the runner, the magic wielder. I had plenty of other skills. But now, I longed to be someone else. Someone who could take advantage of this moment, cut these men down, and free myself.

Instead, I was helpless, just as I had been helpless against the illness that took bite after bite of everything I loved.

I couldn't fight. I couldn't run.

So I started unbuttoning my dress.

I made it three buttons down when I heard a strange sound behind me, like a great unnatural rustling of air. A shadow fell over the streak of moonlight that illuminated Filip's face.

His eyes went wide.

I started to turn around, but before I could, a blur of movement swept from behind me. Something warm spattered over my face.

Before me, a sword impaled Filip's chest. I took in the image of him standing there — eyes wide, like he hadn't yet realized what had happened to him—for only a split second, before chaos erupted.

I stumbled backwards. I couldn't see anything—in the darkness, I just saw limbs and movement and chaos. I tried to seize the chance to get away, but the bearded man grabbed me.

"Back off!" he called out, into the night. "I'll kill her!"

His voice shook.

The figure, who until now had been a smear of shadow, turned.

Vale.

At first, I thought I was hallucinating—from exhaustion, or the blow to my head, or both.

But no. Unmistakable. It was him.

And gods, he was a monster. I now understood why people whispered of him the way they did. This was what I had been expecting to see that first time I met him—a shepherd of death itself. He looked like he had come very quickly, his clothing thrown on hurriedly, his hair messy and unbound and now whipping about his face.

And his wings... they were incredible.

They were fair, which I hadn't been expecting—silvery-white, ghostlike in the night. Even in this moment, I wished I could examine them, appreciate them for the marvel of engineering that they were.

Vale took in my captor, face cold.

My eyes fell to Filip's body, bleeding out on the ground. His hand twitched, reaching up—reaching for his friend.

I felt ill.

Vale lunged.

Pain erupted through my shoulder. I hit the ground so hard I heard something crunch.

I couldn't move. I tried to push myself up and couldn't.

A heavy weight fell to the ground beside me. My attacker's bloody, vacant face stared into mine. Behind him, I could make out only blurry shapes—the white of Vale's wings, the red of blood, and the shadowy silhouettes of body after body hitting the ground.

Wait, I tried to say. *Stop.*

But I couldn't speak.

I couldn't move.

The screams of pain faded into a distant din.

I fought hard for my consciousness, fought just as hard as I had been fighting for answers my entire life, but it slipped away from me anyway.

The last thing I felt were strong arms around me, and the strange, weightless sensation of being lifted up... and gods, I must have been hallucinating after all, because I could have sworn I even turned my head once to see the trees so far below me they looked like stalks of broccoli.

What a strange dream, I thought to myself, as it all faded away.

CHAPTER NINE

I was surrounded by softness. Soft and sleek and… something wonderful. I rolled around and felt silk sliding over my skin.

Silk.

I'd never slept in silk.

My eyes opened. My head pounded. My skin was hot and clammy. I struggled to catch my breath. It had been a long time since I felt this way—so weak, so ill.

When I lifted my head, it felt like an iron weight. I forced myself up anyway.

I was in a bed that was literally triple the size of the one that I slept in at home. The sheets were black silk, the bedspread violet velvet. It was dark in here, lit only by a couple of dusty lanterns that looked like they hadn't been used in a very long time. None of this, actually, looked like it had been used in a very long time—the furniture was all fine but mismatched and outdated, assembled from many different decades, none of them within the last fifty years.

I rubbed my eyes. The events of the night felt like a dream.

But they weren't. It had happened, and now I was here.

In Vale's home.

I had been unconscious in a vampire's home.

I touched my neck, just to make sure —

"I promise I did not eat you."

Vale's voice was low and smooth with amusement.

I turned my head too fast. The movement sent the room spinning, and I swallowed vomit.

He stood in the doorway, approaching slowly, hands clasped behind his back. He looked much neater than he had last night, the monster I had seen replaced with the man I had first met. No sign of those stunning wings, either.

"The bandage is my doing," he said. "But the wound under it is not."

I touched my shoulder and winced. Fabric covered what felt like a vicious cut, and I hadn't felt it only because my dizziness overshadowed it.

"He stabbed you," he said, voice flat. "An accident, when he fell. The rats didn't even know how to wield a weapon."

He spoke with an air of disgust.

I remembered him tearing down those men. The dead face of my attacker nose-to-nose with me. I felt nauseous.

"You killed them."

"I rescued you."

He had. I was grateful for that. But I thought of Filip's hand reaching for his friend...

Vale read my face.

"What?" he snapped. "You think I should have let them live?"

I pursed my lips.

...No. No, I didn't want mercy on their behalf. I'd been distraught that I was too weak to end them myself.

I threw back the covers, but Vale crossed the room faster than I thought it was possible for a creature to move, yanking them back in place.

"You're in no condition to be up."

"I have work to do."

"Three days ago you were near death," he bit out. "You will stay right here."

My eyes widened. "*Three days?*"

How had I been out for that long? I couldn't be away from home for—

"Why were you going to undress for those people?" he demanded.

What?

I shook away that strange question. "I need to go. I have to—"

"You aren't going anywhere. *Why* were you just going to let them do that to you?"

Everything in the world seemed too loud and too sharp right now, but sharpest of all was Vale's expression—like he had been desperate to ask me this question for the last three days. For a moment, I glimpsed the version of him I'd seen in the forest.

"How was I going to stop them?" I said. "It's just a dress. Just a body. Would you rather I refuse and let them kill me?"

I'd always felt disconnected from my body, like it was a strange vessel that only sometimes cooperated with me. It had been my enemy from birth, after all.

Vale looked appalled.

"You could have—"

"What? What could I have done?"

I posed it like a real question, and his mouth shut. I could see the moment he realized that he didn't have an answer.

I blinked and saw those dead bodies. "Was robbing me a crime punishable by death?"

"Raping you would have been. Killing you would have been."

"They didn't do either of those things."

"*Yet*," he snarled. "I've killed others who deserved it far less."

Oh, I believed him.

I looked down at myself. Only at the discussion of my dress did I realize that I was wearing different clothing now— an inordinately frilly nightgown that looked to be at least a century old and like it had never been worn before.

"You undressed me."

"I thought it was *just a body*," he jeered.

Fair enough. And I'd meant it before. But the idea of Vale seeing me naked... that felt like more.

"I preserved your modesty best I could," he added.

I didn't have time to think about this. "I have work to do," I said, firmly, half to myself.

He gave me a strange look—amused? Curious?

"Does nothing bother you?" he said. "You seem totally unmoved that you almost died."

I didn't tell him that I was always almost dying.

"I don't have time to waste on useless things," I said.

"It was strange to see you in such a state, when I found you. And when I brought you back. So... weak."

A wrinkle formed over his forehead, hinting at confusion.

And that confusion, in turn, confused me.

"Weak?"

"You've seemed... In the time we've known each other, you've seemed infallible."

Infallible.

I burst out laughing before I could stop myself.

"What's funny about that?" he said, offended.

I waved him away.

It was funny, of course, because I could not possibly be any further from infallible. I was the very *definition* of fallible.

I pushed the covers back despite Vale's grip on them. And then I rose too fast and immediately fell.

He caught me before I hit the ground.

"I only let you get up so you could see that," he grumbled. "See? You aren't fit to go anywhere."

"I need to go home."

I tried again to rise, and again, I failed. The hot flush of my skin had nothing to do with Vale's hands on me. The floor seemed like it was, quite literally, tilting.

I stuffed down my frustration. It had been so long since the symptoms had been this bad. I had been so preoccupied with the progression of Mina's illness that I hadn't been paying much attention to the progression of my own.

He scoffed. "Home? You won't be making that trek for at least another week."

Now it was my turn to scoff. "Well, that's ridiculous."

"You can't even stand."

"Let go of me."

"I let you fall once, only to make a point. I'd rather not do it again. You're ill, mouse. Far too ill to travel."

"Of course I'm ill," I snapped. "Everyone is ill."

But he gave me a piercing look, one that made my mouth close.

"But you... you are *very* ill."

Four words, and I heard so much in them. I felt like a light had just been shined directly on the weakness I tried to hide.

But Vale was a creature of death. Should I have been

surprised that he smelled its fingerprints on me? Especially now, as it encroached closer than ever.

"My sister." I allowed myself to lean on his grip as I rose. Even let him steer me back to the bed. "I need to—"

"I can send someone to check on your sister."

My heart went cold. "A vampire?"

Maybe I'd looked a little unsteady at the very thought of that, because his fingers tightened around my upper arm as a flicker of annoyance passed over his face.

"What? You're afraid of us, now?"

Only when I thought of one of them getting anywhere near my sister.

But then he seemed to soften slightly. "There's a boy I hire for errands, sometimes. I'll send him. Perfectly human, I promise."

I hesitated.

"I'll have him check on her every day, if I have to," he added, annoyed. "If it'll keep you from wandering out into that forest like an idiot."

A strange emotion passed through me at the irritated urgency of his voice. Why would he care so much?

"Fine," I said at last. "Thank you, my lord."

Vale led me firmly to the bed. "Don't call me that," he grumbled. "I told you. I'm no lord."

CHAPTER TEN

I hated sitting still.

Despised it, actually.

Vale all but threw me back into the bed, and I sat there for half an hour before I was fidgeting, trying to get up only to immediately stumble again. He caught on fast, soon taking watch at my bedside.

"You're self-destructive," he muttered, visibly irritated with me.

"I'm busy."

"You're ill."

So what?

But before I could come up with another protest, he went to the bookcase, withdrew some books, and plopped them heavily onto the bed. "Here. If you want to work so badly, then work."

I picked up the books. They were all written in a language I had never seen before.

"Is this… Obitraen?"

Only at my tone did Vale seem to recognize the flaws of his plan.

"What *are* these, anyway?" I picked up another one of the books and flipped through it. It was illustrated. Graphically so.

My cheeks tightened as I turned the book sideways, taking in a full-page spread. "My, Vale. Your taste is…"

He snatched the book away. "Fine. Then sit here doing nothing."

"What is it, exactly, that you think I *do*, if you thought you could give me a random collection of books written in a language I didn't understand and that would qualify as 'working?'"

His face flushed with something that almost —*almost*— resembled embarrassment. Gods, I wished I could capture that expression. It was a thing of art.

"You're awfully ungrateful of my hospitality," he muttered, turning away.

"Wait."

He stopped at the door and looked back.

No, these books, whatever they were —and I was very sure they had absolutely *nothing* to do with my field —wouldn't help me work. But… still, curiosity nagged at me. How many humans had gotten to read Obitraen books?

"You could read them to me," I said. "If you're just *so* desperate to host."

"Read them to you?"

Was the twinge in his voice disgust? His lip curled as if it was.

"I don't know Obitraen, but you do. If you want me to stay in bed, it would be easier if I had something to do."

Vale thought for a moment, then snatched one of the

books from the bed—not the illustrated one, sadly—and sat in a chair by the window.

"Fine," he huffed. "It isn't as if I don't have much more important things to do, but I'll indulge you if you're bent on being difficult."

"An honor," I said, unable to suppress a smile. "I know you're very busy."

But Vale, for all his supposed busyness and his grumpy reluctance, launched into those stories with all the enthusiasm of a man who would rather be nowhere else in the world.

I LOST myself in those stories. It was too easy. My mind was thick and muddied, and I was exhausted. The first book Vale had picked up was a history book, vampire lore told in short vignettes. Their history was... appalling, but also riveting, every myth and legend woven into a tapestry of blood and betrayal. And yet, even when telling such horrible tales, his voice was smooth and deep, rising and falling like the swells of the ocean. Steady, like a heartbeat or breath. Elegant, like the way his blood looked on the wall.

I didn't remember falling asleep, only that Vale's voice and his stories followed me into my dreams. And I didn't remember lying down or pulling the covers around myself, only that when I awoke, I had been carefully tucked in, silk sheets smoothed tight around my body.

I felt hot and weak, but worlds better than I had before. So I did the only natural thing: I got out of bed and started exploring.

I still couldn't decide if Vale's mansion was the ugliest or most beautiful place I had ever been. Each room I wandered into was more cluttered than the last—an absolute mess, but with the most fascinating objects. I came to the conclusion that Vale must only live in a very small section of this enormous house, because almost every room I peered into seemed to be used as storage. All of them were stunning, but the fourth room made me stop in my tracks, awed.

Weapons. Everywhere, weapons. I was an academic, a farmer's daughter raised in a farmers' town. I'd never held a sword—had barely even seen any, save for those on the hips of city guards. They'd always seemed to be simplistic and brutish instruments to me. Unremarkable.

Not these.

These were works of art. Even I, a woefully untrained eye, could see that immediately. Swords lined the walls, hung straight up and down in slashes of silver and gold along dark wallpaper—swords of every size, every shape, some nearly as tall as me and others light and delicate. On one side of the room, several sets of armor were mounted on wooden frames. Gorgeous, even from a distance—silver metal and black leather and capes of purple silk. Freestanding racks, haphazardly arranged about the space, held axes, bows, arrows.

A few, I realized as my eyes adjusted, were marked with spatters of black.

And there, on an end table right within the door, was a rapier stained with dried red blood—dried, *fresh* red blood. Perhaps from only a few days ago.

The hairs prickled on the back of my neck. The beauty of it all collided with the realization that dozens—hundreds, maybe thousands—had almost certainly been killed with the instruments that surrounded me now.

"You're very bad at resting."

I jumped and almost fell into a rack of arrows before

Vale's hand snaked out to catch me. He pulled me upright, but didn't let me go. Our bodies were close. His eyes were slightly narrowed, searching my face, and I struggled to decode the complexities of what lay within them.

Annoyance, yes—that I expected. But something else, too, like he was waiting for the answer to a question and was nervous about what it might be.

"So I take it you're feeling better," he said.

"Yes. Better." I cleared my throat and pulled away. Then looked to the room.

"You shouldn't be in here," he said.

"How did you get all of this?"

"I take my field seriously, just as you do yours."

"And that field is...?"

"I was a general."

"A good one?"

Even as the question left my lips, I knew it was a stupid one. I'd seen Vale fight now. Like it was an art.

"The third best in the House of Night," he replied, very seriously, and that—well, I wasn't expecting that kind of honesty.

"The first two must have been something to behold, now that I've seen you in action."

The corner of his mouth lifted. "They were. But they are dead, and I'm still here."

And if anything startled me more than his first answer had, it was this.

Because I recognized something in that tone... something human, something vulnerable. My gaze flicked to him, and he was staring at the weapons with an odd, faraway look in his eye. The kind of expression I saw on the faces of those who walked by their family's grave sites.

"You said you oversaw the loss of a war," I said.

He flinched—actually flinched.

"Yes."

"And that's why you came here."

"Yes."

"Why did you keep all of this?"

"It's valuable. I wouldn't leave it behind."

It was more than that. Perhaps the long, hard stare I gave him told him I knew it, too.

"They're mine," he said, after a moment. "If I sold them or left them in Obitraes, they would have been used in someone else's war. Maybe they would have been used against the same men I led. I wouldn't let that happen."

Strange, how vampires and humans were so different and yet so much alike.

"Did you save them because you thought you might need them again?"

A long, long silence. Vale's eyes went distant, his body still —I had never seen a creature who could be so, so still, as if he didn't even breathe.

"No," he said, at last, and closed the door.

Then he turned to me and said, "Do you need my blood, if you're going to insist on wandering around?"

VALE'S SKIN WAS WARM. I felt like I noticed a new thing about it every time I touched him. Even his veins were more elegant than those of a human, the pattern to them more delicate and intentional, the darkness visible in streaks of color like embroidery under the thinnest skin of his inner wrist.

We sat in silence as I took the first vial of blood.

He looked past me, and I followed his gaze to the vase on the coffee table—containing three flowers. I'd given him the last one when I was still half-unconscious, apparently, though I didn't remember doing this. It had gotten a little crumpled in all the excitement of the last few days, but was still just as beautiful as its siblings, petals perfect black and vivid red.

"I still see nothing remarkable about them," he grumbled.

"They're very remarkable. I promise."

"I'm starting to think you might be lying to me."

"And if I am?"

My voice was surprisingly flippant. I was no longer as concerned as I once had been that Vale would, at best, kick me out of his house, and at worst, eat me. Maybe I even suspected some part of him enjoyed having me here.

I watched his blood fill the glass vial. But I could feel his eyes on me, steady and sharp.

"There would be consequences."

Something in his voice made me pause. It wasn't a threat. It wasn't a joke, either, though it held the sweet lilt of one. I could feel his stare on me, and I knew before I looked up the expression that would be on his face.

I didn't move my hands, but the sensation of his skin against mine was suddenly overwhelming.

I met his gaze. The expression was just as I'd imagined it —the faint smirk, the cool stare. And yet… something a little less removed flickered in his eyes as they lowered slightly. Lowered, I realized, to my mouth.

"Consequences," I scoffed.

"What? I'm a dangerous man. You aren't afraid of what punishment might be?"

Goosebumps rose to the surface of my skin, coaxed by the mocking melody of his voice over the word, drawn out slow.

Even I knew that what he was teasing me with, what he

was promising me, was something very different than what he'd done to my attackers in the forest.

Maybe just as dangerous, though.

When my eyes met his, I found it impossible to pull them away. My fingertips tingled, acutely aware of every cord of muscle beneath my hands. My heart beat a little faster. I knew he smelled it.

He had invited me to his bed once before. I'd been tempted then. I'd be lying if I said I wasn't tempted now. Curiosity was my greatest vice.

I'd spent a lot of time over these last weeks thinking about Vale. I was obsessed with him in some ways. I spent all day every day looking at his blood. Admiring its beauty. Admiring that it moved with the same ageless grace as the rest of him.

He was, I'd admit to myself, a very handsome man.

He leaned forward, just slightly.

"Tell me, mouse—"

BANG.

BANG.

BANG.

I jumped. The needle jolted from Vale's arm, resulting in a spray of blood over my chest. I knocked down one of the vials with my elbow, and before I even had time to be horrified by it, his hand had snaked out to catch it—a movement so smooth and quick I didn't even see it happen until he was handing me the vial.

"Cork that. Apparently my blood is valuable stuff."

BANG BANG BANG.

The knocks grew more persistent. Vale looked over his shoulder, into the main hallway and the front door beyond.

I put away the vials, a little flustered.

My first thought was that someone found out what Vale had done to defend me and was coming for revenge. But though the knocks were loud, they weren't frantic or angry. And Vale didn't look concerned, only irritated.

He didn't move.

"Do you... want to get that?" I asked.

"No, I don't."

I'd forgotten. He didn't like answering his doors.

But the *BANG*s continued, a rhythmic beat growing steadily faster, until Vale finally let out an aggravated sigh, rose, and went to the door.

I followed him. I couldn't help it. He was right; I was nosy.

Vale opened the door with a single abrupt movement.

I stumbled backwards.

The person at his doorstep had no face.

CHAPTER ELEVEN

The faceless person stood there silently.

Or maybe it wasn't a person at all—just the suggestion of one. He—*she? It?*—was only a silvery outline in silhouette, the edges of its form streaks of painted moonlight, and the core of its body nearly clear. I could see the forest straight through the center of its chest—straight through the center of its face. It was nearly as tall as Vale, though willowy, its limbs thin and slightly formless, only a suggestion of bones and muscle.

Vale looked totally unmoved.

"I told you not to come back here," he snapped.

If the form was capable of either hearing or understanding him, it showed no sign. Instead, it simply held out a hand. A single letter sat in its palm.

"I don't want that," said Vale.

The form did not move.

Vale groaned and snatched the letter away.

"Fine. There. Now go."

The figure started to fade, and I watched wide-eyed, eager

to see how it would leave. But Vale just slammed the door shut, and the look on his face made me startle.

He looked... irritated. More than irritated. Irritated was how he had felt with me when I first showed up at his door. This was an even harder expression, his jaw tight, his fist clenched around the letter, now crumpled in his fingers.

"What was that?" I asked.

"Nothing."

"Was that Nyaxia's magic?"

"Was that—what?" He looked at me, blinking, like he'd been so lost in his thoughts he'd forgotten I was there for a moment. "Oh. Yes."

"So that's from your home."

He scoffed. "My home."

"From Obitraes," I clarified.

"Oh, I understood you."

I paused. "Well, you're upset," I said, mostly to myself.

"I'm—" He stopped short, whirled to me, snapped his jaw shut. "Yes. Yes, I'm upset."

What was I supposed to do? *Not* ask?

"Why?"

"*Why?*"

He turned his back and paced, and I got the distinct impression that he wasn't really talking to me anymore.

"I'm upset because they won't take no for an answer. Because I'm not doing this. I'm not going back to Obitraes. I'm not going to help put some—" His lip curled. "Some nobody on my throne. I'm not going to lead another losing war. I am not going to do any of those things, mouse. Not a single one of them."

I looked to the letter in his hand. Now completely crushed in his fist.

He let out a long breath and straightened. "I'm—I apologize." He seemed a little embarrassed. But he shouldn't have been. I didn't mind seeing him with his guard down.

"Is that what that says?" I said. "They're asking you to return to the House of Night?"

"Yes, and it doesn't seem to matter that I've told them no, many times over."

"So, why do they keep asking?"

He let out a light scoff. "Because no one else would help."

"Because the top two generals in the House of Night are dead."

Vale blinked, mouth tightening with an almost-smile. "Yes. But just as well, because those bastards wouldn't have helped them, either."

"Who's... them?"

"No one worth talking about."

"But why do they need you?"

I assembled the pieces of our previous conversations and my sparse knowledge of Obitraen history.

"You're Rishan," I said. "And the Hiaj are in power now. Does that"—I nodded to the letter—"mean that there might be a change?"

The expression of surprise on Vale's face was unmistakable.

Confirmation.

"Your people are attempting to retake the throne." I was pleased with myself for putting this together, the same way I was satisfied when I solved a difficult equation. "And they're asking you to come back and—"

"And help them lose a war," Vale snapped. "All in the name of some bastard king."

I had never seen him like this. He looked like he was crawling out of his skin.

"You don't like this man," I said. "Or, uh... woman. Person."

Who was I to make assumptions?

"He's... not king material."

"You've met him?"

"A long time ago, yes."

"And you didn't like him?"

"I—" He seemed to be at a loss for words. "I wouldn't bow to him. *No one* would bow to him."

I stared blankly at him.

"What?" he snapped. "You look as if you're about to tell me I'm wrong, so go ahead. Do it."

"Right now, your people are not in power. Is that right?"

"Yes."

"And what does that mean for them?"

A muscle feathered in Vale's neck. He didn't answer right away.

"Are the Hiaj fair rulers?" I asked.

He let out a short scoff. "*Fair*. Of course not."

An unpleasant, unflattering understanding settled over me. My lips thinned. My mouth tasted sour, like it always did when rude words I shouldn't say were lying in wait.

I said, curtly, "We should finish our work."

I started to turn, but Vale caught my shoulder.

"Say what you're going to say, mouse."

"I wasn't going to say anything."

"Don't lie to me."

I stared at him, unblinking. I didn't know what to make of the way he was looking at me—like he actually wanted to hear my opinion.

Or thought he did.

Keep your mouth shut, I told myself—but I'd never been good at listening to my reasonable voice. He'd jabbed at something I tried to hide, a frustration that now surged faster than I could stop it, and I wasn't even sure why.

"It's just... something being difficult is not a good reason not to do it."

He pulled back, offended. "It isn't about it being *difficult*."

I tried to hide my skepticism and apparently failed.

"*It isn't,*" he said. "It's about principle."

"Principle?" I choked out a humorless laugh. "Your people are asking you for help and you're refusing because of principle?"

"It's just not the way things —"

"My sister is dying."

I blurted out the words in a single rough breath.

"My sister is dying and my *whole town* is dying, Vale. And everyone else thinks that we can hope or pray or dream our way out of it. They're just like you. They're refusing to seek better answers because of *principle*. Because it's *just not done*. And every second they waste time waiting for a stupid dream is another lost life. That is someone who is the most important person in the world to another, somewhere."

Vale didn't blink. And I didn't know why, but I couldn't stop talking. The words just poured out of me.

"I know what it feels like to be *helpless*," I ground out. "*You* don't. *You* don't know what it feels like to be surrounded by five men and know you can't stop them from hurting you. *You* don't know what it feels like to see the people you've grown up with wither and die. You —"

You don't know what it feels like to watch yourself die.

I stumbled over that one.

"And I can't blame anyone for bad luck and misfortune," I said. "But if I ever knew that someone had a chance to help them — had a chance to save even one of those lives and *denied* it —"

I blinked and saw my sister, slowly grinding away into dust—my lively sister who was everything I was not, who was life when I had always been death, who was warmth when I had always been cold. My beautiful sister who deserved to thrive so much more than I did.

I hadn't stopped to breathe. When I did, it was a jagged, ugly sound, broken with an almost-sob.

Vale had gripped my shoulder. His thumb rubbed my skin, right at the boundary of the neckline of my dress. Something about that touch steadied me. It was a comfort, a reassurance, and a question.

My face was hot with embarrassment. I shouldn't have said any of that. It was uncalled for.

Vale's other hand came to my cheek, and when he pulled back, his fingers were wet. He looked down at that for a moment —my tears on his knuckles—then back at me. I straightened and stepped away from him. I felt unsteady. Drained.

He was calm now, too. Just looking at me. Thoughtful.

"I'm sorry—" I started.

But he said, "I want you to show me my blood."

I DID AS HE ASKED. We had to go into three different rooms before we finally found one with a wall clear enough for my instruments. I blew out all the candles and set up my lens. A part of me didn't even want to risk using it here—they got expensive after awhile, and if this one broke too, I'd really have to scramble for the money for another—but it seemed important to grant Vale's request.

I wanted him to see in himself what I saw of him every day. The beauty of it. The miracle of it.

When his blood bloomed to life over the wall, I drew in that same little inhale. I did it every single time.

Vale's expression was utterly still, save for a very, very slight widening of his eyes. He slowly leaned forward to rest his forearms on his knees.

"So this is it."

"This is it."

"Why does it look like that? The dots?"

"That's your blood at its basest level. Very, very small."

He made a low, unconvinced sound in his throat.

"And what is special about it? Different?"

I rose and went to the wall, examining his blood up close like I had so many times before. "See how it moves? It's different than human blood. The color, too. The shape. It deteriorates differently." He didn't speak, didn't stop me, so I found myself slipping into my own enthusiasm—explaining to him all the ways his blood differed from that of humans, all the little ways the magic of his nature and his goddess imbued it. All the ways it defied death.

Afterwards, he was silent for a long time. "You believe this," he said, at last. "That it could help."

"Yes. Yes, I do."

"Vampire blood has never helped anything."

I looked back to the projection on the wall. I needed to take it down, and fast. The machine would start smoking at any moment. But I touched the wall—touched the curve of each flower-petal shape.

"Your blood is..."

Gods, it was so many things.

I settled on, "It could save us."

I was lost there, in that projection, until Vale said, "That's not true."

I turned back to him. He didn't look at the blood. Only me.

"You," he said. "You are saving them."

He said it with such conviction, such certainty, that I did not know how to respond. He rose, hands clasped behind his back.

"Whatever you need," he said. "My blood. My books. My knowledge. Anything. It is yours."

CHAPTER TWELVE

I had, apparently, piqued Vale's interest, because from that moment forward, he wanted to study with me all the time. We dug through his libraries and studies together, and he helped me find books that might be relevant to my work, then translated them for me as I scribbled frantic notes in my notebooks. Time blurred together, every minute morphing to hours until my head started to bob over my books and Vale would force me to rest.

"Is this how you live?" he asked, appalled, to which I blinked blearily at him.

"I have work to do," I answered, because this was obvious. To that, he snorted and scoffed and dragged me off to bed, and then sat there to make sure I stayed—because I'd been foolish the first time and let him catch me sneaking out.

I couldn't help it. There was so much knowledge in Vale's house—so much to learn. I wanted all of it. I wanted lifetimes, eternities, to absorb everything that he knew—to experience the world as he had.

Two more days passed, then three. My health improved. I

toyed with the thought of leaving on the third day, but Vale said, very seriously, "You still are unwell. You're in no state to travel."

And later, I would lie in bed and swallow shame, because I could have argued with him — *should have* argued with him.

But I didn't want to.

Because maybe some part of me found a strange kinship with him in those exhausted, sleep-deprived days. I'd watch him read his Obitraen books to me, watch something flicker to life over his face, a fervent curiosity that mirrored what I so often felt and always dampened.

I had thought nothing could possibly be more beautiful than Vale's blood. I had been wrong.

And when the days passed, and my exhaustion and my enthusiasm led me to loosen my typically-closely-held control over my socially unacceptable attitudes, my raw enthusiasm leaking through as I talked excitedly to Vale about some theory or another, I turned to see him staring at me, brows drawn. His expression made me freeze, my face flushing — because I'd let down a wall I shouldn't have and wasn't sure what I might have revealed beyond it.

"I—" I started.

But he just said, calmly, "You are a very beautiful woman."

It wasn't an invitation, like the first night he had asked me if I wanted to spend the night with him. He wasn't flirting with me. No, it was an observation, clear and simple as those in the books spread before us, and Vale simply let it lie there and then turned back to his book.

CHAPTER THIRTEEN

I needed to go home.

I knew it the moment I opened my eyes that day. The thought came with a sharp stab of guilt, like a haze had been cleared and I realized all at once what I had been doing.

I had been with Vale for a week. A week, in a world where time was so precious and cruel.

I needed to go home.

I told Vale this and didn't know what to make of his slow nod and quiet demeanor. He insisted on sending me home on a magnificent black horse—a horse that was probably worth more than all of my belongings put together. "You aren't well enough to walk all that way," he said, when I tried to protest.

He helped me mount it, which I wasn't expecting, his hands firm and big around my waist. His grip sent a trill up my spine that struck me in places I wasn't expecting. When I was seated and he stood beside the horse, his hand still casually resting on my thigh, that touch was the only thing I could think about.

"Thank you for the hospitality, Vale."

He shrugged a little, as if he was trying hard to make it seem like a great inconvenience.

Still, he didn't move, and I wasn't sure why. That hand still rested there, right on my thigh.

Was he waiting for me to say something? Had I missed a cue that I should know? I did that often. I looked down at that hand.

"What—"

"May I write to you?" he asked.

My mouth closed. I blinked at him.

"May I write?" He sounded vaguely irritated, and I wasn't sure why.

"Yes," I said, at last. "Of course."

There was nothing *of course* about it. It was unwise to allow Vale to write to me. Unwise to allow more evidence of Nyaxia's cursed children into my house, where it might draw even more ire from the gods than we had already earned.

A voice in the back of my head screamed this to me. A voice that was far too easy to silence.

I had, after all, brought so much of Vale into my home already. His blood. His books. And how I felt as if I was covered in him, right down to my skin. Right down to my heart.

What harm would some letters do?

He exhaled, shoulders lowering. The irritation faded. I realized that maybe he hadn't been irritated with me, but with himself.

Relief. He was relieved.

And the truth was, so was I, because the thought of leaving Vale—the thought of being able to continue what we had started together this last week—

Vale stepped away from the horse. His hand was the last part of him to move. I watched it leave my thigh.

"Travel safely," he said.

I gave him a small smile. "I'll see you in a month, Vale."

And he returned that smile—a thing so lovely I barely even noticed the teeth. "I'll see you in a month, mouse."

WHEN I ARRIVED HOME, Mina threw herself at me. "I missed you! I was so worried about you. They said you were fine, but I didn't believe them. You were—well, you were—"

I didn't mean to stiffen under her embrace, but I did anyway. Not at first—at first, it was a welcome reminder that she was safe, that I had made it home. But then I just felt everything else. I was always so acutely aware of sensations and textures, and now I felt Mina's frailty most of all.

She felt it and pulled away, brows drawn. Hurt.

I looked down at myself. A fine coating of white-grey coated my clothing where her skin had touched me.

"He didn't hurt you, did he?" she asked. "I was so worried, Lilith. I was—I was so, *so* worried."

I swallowed a stab of guilt.

She was worried, and I was... I was...

I was *happy* there. In no great rush to come back. No great hurry to escape the quiet comfort of Vale's home.

The final remnants of the dream I'd been living in for the last week faded away.

I hadn't even written to her. What kind of a sister was I? Too preoccupied with—with some man—

"He didn't hurt me," I said. "He was..."

Kind. Caring.

I settled on, "He let me recover there."

Her mouth pinched. "When you were bleeding? You're lucky you made it out of there alive."

I felt foolish for not putting that together sooner—that I had been bleeding, and I probably had been very, very tempting to Vale.

"He showed no interest in eating me," I chuckled. "Don't worry."

And yet, as I said it, I heard his voice: *You are a very beautiful woman.*

Felt his hand on my leg.

Mina was giving me a strange look.

"Well. I'm glad you're alright. I was... we were just all so worried about you, alright? Don't you dare leave me like that again."

I agreed, but it was a lie. That was the cruel joke with Mina and I. She'd leave me, or I left her. I'd do everything I could to make sure it was the latter.

"A LETTER CAME for you this morning," Mina told me later that evening. "It's in your office. It's... strange."

She was right. The letter *was* strange. But strange in a way I now was beginning to know quite well. The paper of the envelope looked as if it could be a decade old, yellowed and a little crumpled. It was closed with a red wax seal.

I knew right away that it was him. I smiled to myself when

I held it, just because it reminded me so much of him. It was so... well, so vampiric.

I opened the envelope. Inside, there were a few torn-out pages of books with notes and translations scribbled in the margins in the handwriting I now recognized as Vale's.

And then there was a letter. At the top was my name, and then several black drips of ink, like he'd hovered the pen over the page for a long time, thinking about what to write.

Lilith,

I hope you made it home safely. I found some more notes for you.
I thought you wouldn't want to waste the time without them.
I welcome any letters you wish to send before your visit.
I will help you however I can.
If you want it.
Vale.

I didn't realize I was smiling until my cheeks began to ache.

It was so...

Familiar. So strangely familiar. Just a few stilted lines. None of the flowery language of polite society.

And yet, I knew it said so much that wasn't written in these words, too.

I set the letter down and jumped when I realized Mina was standing behind me. I cursed and shoved the letter into my pocket, even though I didn't know why my impulse was to hide it.

But she had seen it, anyway.

"You startled me," I said.

"Be careful, Lilith," she said. "You know what will happen if they know. If they find out."

My mouth was dry.

"I don't know what you're talking about."

I didn't want Mina to know, either. But who was I fooling? She was so much smarter than anyone ever gave her credit for.

And she was smart enough to know when I was lying.

She gave me a hard look. "Be careful."

PART FOUR

THE FOURTH ROSE

CHAPTER FOURTEEN

I wrote to Vale every few days, and then every two days, and then every day. Sometimes, even, multiple times a day.

Ravens would appear in my garden, ready to deposit his latest letter or take mine back to him. Sometimes he sent his messages with magic, the parchment appearing in little puffs of white-blue smoke—those letters were always his most frantic, like he'd had an idea he couldn't wait for a raven to tell me about, and I'd be lying if I said I didn't devour those the quickest of all.

Vale's enthusiasm was impressive, but even more surprisingly, it was… familiar. Before, I had respected him, the way one needs to respect a great beast by recognizing that it's something older and stronger and more powerful than you. But with each one of these letters, that respect turned from a respect of nature to the respect of a friend.

His handwriting was sometimes sloppy, his notes scrawled in margins or at an askew angle across the parchment, like he was in such a rush he couldn't stop long enough to straighten the paper. I could imagine him writing them, leaning over a messy desk, hair falling around his face, surrounded by open books. He had less reverence for the artifacts around him than I did—he had no qualms about tearing out pages of books to send to me, folded up and scribbled on.

When I had first met him, it had been impossible to imagine him embodying that kind of enthusiasm. But now I could so clearly picture him as the general—the general attacking problems with strategic, unrelenting verve. He had never been a man of science, and his inexperience showed, yes —but he also learned fast, and he wasn't afraid to ask questions or admit his own ignorance, a quality that many men lacked. Much of the information he sent me was genuinely helpful, and when it wasn't, he wanted to learn why.

It wasn't just work, either. He wove little fragments of his life into those letters, too, doodled in the corners or at the bottom of the page. A little drawing of a bird he'd seen on his balcony railing. Mundane observations about the weather: *The wind is cold today. How can you people call this spring?*

But I liked those things, too. I liked that they so easily allowed me to imagine him, shivering a little under the night-time breeze. I even liked that he wanted those banal details from me, too.

One day, he ended his letter with a drawing of a nightbane flower, and a tiny note beside it: *sweet with a bitter bite.*

It was an afterthought, like he hadn't even known that he'd drawn it. The rest of the parchment was filled with information he'd taken from his Obitraen books—useful stuff, actually, far more useful than a little flirtatious drawing.

And yet, I couldn't tear my eyes away from that flower. From his words beside them. Those letters were not scribbled. They were delicate and soft and elegant, like he had been very careful about how his pen had caressed them.

Sweet with a bitter bite. I could still feel the way his breath had skittered over my skin when he said those words to me that night, when he told me he thought it was what I would taste like.

And sometimes, in the rare moments I allowed myself to sleep, I would lie awake staring at the ceiling, eternally conscious of the way my clothing felt against my skin. And I would slide my own fingertips over my inner thighs—higher —and imagine, without meaning to, what his caress would feel like there, too.

Good, I decided.

It would feel good.

THE TRUTH WAS, I was shamefully, secretly grateful for the distraction of my task and Vale's letters. Because I worked, and Mina withered.

Every morning I swept the dust from the door. Every evening it was covered over again. Church hymns rang through the streets, the air thick with the smoke of another funeral pyre, and another, and another. The smoke was thinner each time, because now, there was often so little left to burn.

I forced myself not to think about what Mina's pyre would smell like. I told myself I wouldn't have to find out.

Mina and I did not discuss her decline. What was there to say?

But the blood drained from my face the first time I came home to see Thomassen sitting at our kitchen table, his hand in Mina's, their heads bowed.

An acolyte of Vitarus in my house—the same house in which I had an entire room dedicated to the blood and belongings of my vampire... friend. Dangerous.

But what frightened me more were the silent tears that rolled down my sister's cheeks, because I sensed what this was the moment I walked into the room.

I had long ago come to terms with my own cruel mortality. But it isn't easy to accept that kind of ugly truth. I went through my struggle when I grew old enough to understand what death meant. In the years since, I'd watched so many others go through it, as their eyes grew hollow, their skin dusty. I saw the desperation as they looked up at the sky, where maybe somewhere the god that cursed them lurked, and I knew they would do anything, *anything* for more time everyone knew they wouldn't get.

When I came home and saw the priest holding my sister's hands, I knew that, for the first time, Mina felt that desperation.

That terrified me.

My sister had looked up and given me a weak smile.

"Sit with us," she said.

In the same tone of voice that she had asked, *Stay.*

Stay, I wanted to beg her.

But no, I wouldn't pray to the god that had damned her. I wouldn't help her come to terms with a death I refused to let her meet.

"I can't," I said, and went to my office without another word. I didn't stop working until dawn, and even then, I fell asleep over my books.

But Thomassen came more and more often, and death crept closer and closer.

If I was less distracted by my work and the grief I tried so hard to stave off, perhaps I would have been more concerned by the acolyte's constant presence in this house. Perhaps I would have given more thought to the way he watched me, the lingering stares at the doors I left ajar.

But I was used to being judged—too used to it to realize when judgment became dangerous.

I didn't have time to worry about one old man's thoughts about me. I had to work.

I was running out of time.

BUT THEN, one day, when nearly a month had passed since my last visit to Vale, something shifted. I slept in my study that day, as I so often did now, and I woke up to a pile of Vale's letters, strewn across my desk. Four of them, in the sparse hours I'd been asleep.

My heart jumped with either anticipation or dread. So many in such a short span could only signal something wonderful or terrible.

It turned out it was the former.

Vale had made a discovery. I flipped through his letters—pages upon pages torn from one of his books. I'd gotten used to his scrawled handwriting, but the translations in the margins were even messier than usual, as if he'd been writing so fast he couldn't even stop to form real letters. It took me hours to fully decode them.

When I did, I gasped.

He had found a crucial missing piece. The text was old, detailing experiments done on vampire blood in Obitraes. Yet, despite their age, the figures answered so many of the questions I had been grappling with about how to effectively distill vampire blood into something different. Vale and I hadn't found much in the way of Obitraen science—vampire society, it seemed, was much more inclined to work with magic instead.

But this... it was exactly the sort of information I'd barely allowed myself to dream of.

"Vale," I breathed under my breath. "Vale, you—you—"

I was grinning so widely my cheeks hurt. I probably looked like a lunatic, half-mad with exhaustion and hope. I hadn't changed my clothes in days, and I figured another day wouldn't do any harm, because I launched myself right back into work.

Hours blurred into days. New equations became new formulas became new vials of experimental potions. Vials of experimental potions became tests as I gave them to my ailing rats.

And tests became medicine as those rats grew less and less sick.

The next batch, too. And the next.

And then, one bleary morning, I found myself standing before an entire cage of healthy, active rodents, cradling those vials in my hands like a newborn infant—and medicine became a cure.

A cure.

It was only fitting, of course, that this was when everything fell apart.

CHAPTER FIFTEEN

I opened the door, and Farrow stood there, his sandy hair wild and eyes wide. Sheer terror.

He sagged against the frame when I opened the door, like he was so relieved to see me that all his muscles gave out.

Mine, on the other hand, tensed, as my fragile newfound hope smashed to the floor.

"You have to go."

He said it so fast that the four words ran together in a single exhale.

"What—"

"They're coming for you," he blurted out. "They came to the city looking for help. They're coming for him, and then for you. You have to go."

He grabbed my arm, as if ready to haul me away by force. But I remained rooted, stuck, dread falling over me like a cold shadow.

I didn't need to ask who "they" were.

Because I could picture Thomassen's cold, suspicious

stare. I could picture Vale's ravens and magic. I could picture all the little marks of my friend I left around this house, now so blatantly, foolishly, *stupidly* obvious.

What was the obvious end to this story? Ignorant zealots who didn't want to die were presented with a god that no longer loved them, and an illness that just kept spreading, and a vampire upon which they could blame it all.

Easy. A simple equation.

They're coming for you.

They're coming for him.

"You have some time, but you need to leave," Farrow was saying in the background. "You can stay in my apartment in the city. I'll have a carriage waiting and—"

"No."

I wrenched my arm out of his grip, turning back to my office.

"No?" he echoed.

"Take Mina and leave without me."

"But Lilith—"

Farrow kept talking, quickly, but I wasn't listening to whatever he was saying. I let his voice run into the background.

We had no time for words. Only actions.

I grabbed my coat. My bag. My precious, precious bag.

Mina. I needed to—

"What do you mean, no?"

Funny, how Farrow's voice disappeared into the din of my rushing blood, while Mina's, weak as it was, made every other sound disappear.

I could count on one hand the number of times I'd heard her sound like that. Enraged.

I turned slowly. She stood in the doorway. Or, maybe "stood" was too strong a term—she leaned heavily against the frame. I was struck all over again by how weak she looked—it seemed like she had even shrunk. How long had she been standing there? Only long enough to hear Farrow arrive, and yet dust already gathered in the ridges of the floorboards at her feet.

I realized, with a sinking feeling, that Mina couldn't go anywhere, no matter what Farrow said.

We were running out of time. My sister's was almost gone.

My eyes slipped away. I rummaged through my bag.

The medicine. It was early. It was risky, but—

"What do you mean, *no?*" she repeated. "Where are you going?"

"I just…" My tongue wouldn't cooperate with me.

She made a strangled sound, almost a humorless laugh. "You're going to him."

If I hadn't been so distracted, I might have been surprised. My sister saw more of me than I thought she did.

I just said, "I have to go. Here—"

"Enough, Lilith. Just—just *stop.*"

Mina's voice cut through the air like a blade, sharp enough to make me pause.

"Look at me," she demanded.

My fingers, deep in my bag, closed around that single precious vial of medicine. I couldn't bring myself to lift my eyes.

"*Look at me.* You never look at me anymore."

I turned around slowly.

I never found it necessary to look people in the eye when I spoke to them, a bad habit since childhood. But with Mina… it was different. It wasn't about discomfort or disinterest or manners. I had to force myself to meet her gaze, to acknowledge all the blatant signs of death devouring her. She stepped closer, not blinking. She had our father's eyes. Light and bright, like the sky.

Right now, they begged me for something.

My risk calculation resolved to a single solution.

"Give me your hand," I said.

It wasn't what Mina wanted from me. I knew that. But I couldn't give her that warmth, that affection. What I could do was try to save her life.

"Don't go there," she said. "We can fix this."

Ridiculous. What would "fixing it" look like, in her mind? Restoring the status quo? Curling up to die quietly in a socially acceptable manner?

No.

"I *am* fixing it," I snapped. "Give me your arm."

"This isn't—"

"I refuse to let you all die." I didn't mean to shout. I did anyway. "It isn't supposed to take you and *I won't let it*. So give me your gods-damned hand."

Her jaw tightened until it trembled. Those blue eyes shone with tears.

But she thrust out her hand, exposing a forearm of pale skin so thin the webs of veins beneath were easy to see.

I didn't give myself time to doubt as I filled the needle and injected her. She winced, and I realized I was so used to the durability of Vale's skin that I'd pushed too hard. A veil of dust fell to the floor. So fragile now.

I withdrew the needle and turned away abruptly.

"Don't open the door for anyone. I'll be back as soon as I can."

I thought she'd tell me to stay, again. Thought she'd still try to talk me out of it. Farrow was looking at me like I was some kind of foreign beast—the same way he looked at a specimen that didn't make sense, his brow knitted, jaw tight.

He was seeing something new in me, something that didn't reconcile with the version of me he had always known.

Maybe I was seeing that in myself today, too.

I couldn't tell if it was a good or bad thing.

"I'm coming with you," Farrow said.

I didn't look at him. I grabbed the axe from the wall and threw my pack over my shoulder. "Fine," I said. "Then let's go." And I slammed the door behind me.

CHAPTER SIXTEEN

W e galloped hard through the morning. My horse, the one Vale had given me, was strong and fast. Farrow's, however, was not used to running for so long and over such uneven terrain.

"Don't slow for me," Farrow called after me, and I let out a rough, wild laugh that I was grateful he didn't hear. I never planned on slowing for him. I'd ride as fast as I could.

I felt like a fool.

A fool because I had spent all this time worried about the dangers my relationship with Vale would pose to me, my sister, my town. But it had never occurred to me that *I* would be dangerous to *him*.

Thomassen had gone after Vale with several dozen men, Farrow had told me as we ran—young and strong ones. They'd brought weapons and explosives and fire. And they'd brought the most dangerous things of all: desperation and rage.

The acolytes believed that Vale was the reason for the curse. They'd convinced themselves that slaughtering him, offering his tainted blood to Vitarus, could end the plague. They convinced themselves that they could only save themselves, save their families, through this murder.

It didn't matter that Vale had lived here far longer than the plague had. It didn't matter that we had sacrificed to Vitarus many times before, and it hadn't worked. It didn't matter that they had no evidence that Vitarus even remembered us at all —even remembered he had damned us.

No, logic doesn't matter in the face of fear and emotion. Logic falls to its knees before hatred, and hatred flourishes in fear—and my people were terrified.

I was terrified, too.

I knew Vale's blood so intimately, now. I knew what it would look like spilled over the steps of his home, spattered over the faces of the people who came to kill him. I'd dissected many animals, many cadavers. I knew what Vale would look like with his guts pulled apart.

I raised my eyes to the sky. The sun was now high, beating down on my back and forehead through the tree leaves.

That, I did not know. What would happen to a vampire in daylight. I thought that after all I had seen, known things were the most terrifying. But this—this unknown—made me sick to my stomach.

I smelled the fire before I saw it. Burning flesh—in a plague, one recognizes that scent innately.

Finally, I saw the gates of Vale's estate glint through the tree branches, open and gently-swaying in the breeze.

I kicked my horse and tore through it.

Behind me, Farrow shouted my name, and I ignored him.

Because before me, there was only blood.

CHAPTER SEVENTEEN

ale had fought them.

The house was bleeding. Blood dripped down the white stone face, pouring from a broken window on the second story, where a limp body hung draped over broken glass, a sword dangling from his motionless grip.

Blood painted the front steps of the entryway—smears of it, pools. Handprints on the door, on the handles. Strokes of it ran in rivulets down the pathway, collecting in the spaces between the brick pavers. It sank into the rose bushes. Into the grass.

Was it horrible that I wasn't horrified? Was it horrible that I was relieved?

Because it was all red blood—human blood. Blood that belonged to the lifeless bodies strewn around the property. So many I couldn't count them. A massacre had happened here.

Farrow had told me that Thomassen had come with two dozen men. Surely few of them remained.

Maybe Vale had escaped. Maybe he...

But then, as my horse slowed to a trot beyond the gates, I

saw it: the black blood mixed in with all that red. Smears in the grass, along the path. More of it down the path to the back of the house.

Too much of it.

I kicked my horse into a run toward the back of the house, ignoring Farrow's calls after me.

And when I saw him, my heart sank and leapt at the same time.

For some reason, the phrase that flew through my mind was, *Vale.*

My Vale.

Only a handful of men remained alive, but Vale was so injured that he wasn't fighting anymore. They had dragged him outside. He was on his knees in the garden, white and red flower petals around him. His head was bowed, black hair covering his face. His wings were out, the white feathers gorgeous in the daylight sun—gruesome contrast to the spatters of black blood and the open burn sores spreading across them.

He looked up as I approached, revealing a face mottled with blackened burns.

His eyes widened.

I didn't even let my horse stop before I was dismounting, running, running—

I threw myself over Vale, tumbling to my knees before Thomassen.

"Stop! Enough!"

The world stopped. The priest, and the four men behind him, leaned back a little, like they had to take a moment to figure out if I was really here.

A rough touch folded around my wrist from behind. Concern. Restraint. It said so much.

"Mouse..." Vale rasped.

His voice sounded so hollow. It reminded me of Mina's. Close to death.

I didn't look at him, though I was so acutely aware of his form behind me, the faint warmth of his body where my back was only inches from him.

Instead I met Thomassen's gaze and refused to relinquish it. The acolyte wasn't injured, though blood smeared his robes. Had he stood back and let the others do all the fighting? Waited until they wore Vale down enough to step in and make the final blow?

"Stop this insanity," I said.

His confusion fell away in favor of hatred again. He gripped his sword, eyes briefly falling to my axe—gods, did it even count as an axe? It was barely more than a hatchet—before returning to my face.

"Step away, child," he said. "Don't do anything foolish."

"If you kill him, then you're killing all of us."

The priest scoffed, lip curling. "We should have done it the moment the plague began. Perhaps a sacrifice of one of the heretic goddess Nyaxia's children would have been enough to end it. Maybe it would have been enough to appease Vitarus."

I wanted to laugh at his foolishness. I wanted to scream at his ignorance.

"Why is it so difficult for you to understand that Vitarus doesn't care about us?" I spat. "He has taken a thousand lives from us. Ten thousand. And that hasn't been enough to appease him. Why would this one be any different?"

"You're not a stupid girl," the priest sneered. "A strange one, but not a stupid one. You know why. Because of what *he* is." He jabbed his sword toward Vale. "Because of who he worships. Because of the goddess who created him. Look around you. How many of your brethren has he killed? And you expect us to let him live?"

I looked into the eyes of the men around him, and I didn't see brethren. I saw people driven to ignorance and hatred. I saw people who were willing to kill whatever they didn't understand just for a chance of a chance that it would help them.

Nothing would stop them from killing Vale.

They would happily kill me, the strange spinster woman that never had laughed at their jokes or indulged their mindless conversations, to get to him.

I liked solving problems. But I was now stuck in a conclusion decades in the making, helpless.

Behind me, Vale's breaths were ragged and weak. I would have thought that he wasn't even conscious, were it not for his grip on my wrist, still strong, even as his blood dripped down my hand.

"Please, Thomassen. Please. I—" My voice caught in my throat. Cracked. "I need him."

The words tasted thick. Heavy. They seemed to sit in the air. I could feel their eyes on me, on Vale, on me again, the way my own often darted between pieces of an equation, and I didn't like the answer they were drawing.

"He could be the cure to this," I said, desperate.

Wrong thing.

Realization fell over Thomassen's face. Realization, and then hatred.

"I defended you," he snarled. "When they talked about you. About your father. About your family. I defended you, child, from horrors you don't even understand. But I was wrong. You'll only spread this further."

He lifted his sword.

Everything went too slow and too fast at once.

Behind me, Vale tensed, pulling me back.

I yanked my hand from his grip, rising.

It was like I was outside my body, watching someone else lift that stupid little axe—watching someone else swing it. I was a scientist, not a soldier. My swing was clumsy, but I threw all the strength I had into it.

Hot blood spattered across my face.

Numb, I pulled the axe from Thomassen's shoulder. I stumbled backwards a little—it was hard to get the blade from the flesh.

Shoulder. Not deadly. Try again.

I swung again, this time for the throat.

It's an interesting sound that one makes when they're drowning in their own blood. No scream, just a gargle and the empty hiss of air. Wet, weak death.

I had moved fast, for all my inexperience. It took a few seconds for the other men to realize what was happening. The priest staggered.

I felt a strange sensation. Something wet over my torso.

Pain, slow.

I looked down to see blood all over my shirt.

Commotion. Noise. It seemed very far away. I looked up and saw familiar sandy-fair hair, a wiry figure yanking a sword from one of the guards as the priest staggered.

The priest's? Or...

I hit the ground hard as a grip from behind shoved me away—Vale. Vale's movements were nothing like the graceful death I'd seen in the forest that night. No, these were lurching, desperate. Survival more than skill. Like a dying animal.

CRUNCH, and a head fell to the ground. One guard, before he could turn on Farrow.

He killed the second with his own sword, torso opened and bloodied over the grass.

Thomassen still stood, somehow... still stood, covered in

blood, a dead man walking. Maybe his god helped him a bit, after all, because he somehow managed to turn—to—

"Vale!" I screamed.

Vale whirled around just in time. Thomassen's sword went through his shoulder.

But Vale didn't flinch.

A terrible damp crunch rang through the air. And when Thomassen's body slumped to the ground, something red was clutched in Vale's hand. It looked like a ball of blood, at first.

Then I realized, after a few seconds of dull blinking...

A heart.

Thunk, as Farrow's sword fell to the grass.

Thump, as Vale let the heart drop beside it.

And then silence.

Birds chirped in the distance. A faint breeze rustled the tree leaves. The scent of spring was so overwhelming, it almost drowned out the scent of blood.

Nothing existed except for Vale and I, our gazes locked. For a long, breathless moment, I couldn't look at anything except for his dark-gold eyes, staring at me through gore-streaked tendrils of hair, through smears of blood.

Then he collapsed.

I leapt to my feet, ignoring the pain of my own injuries, and ran to him. Farrow knelt beside him, too, and started to roll him over to look at his face, but I said, "No! The sun."

Up close, the burns on Vale's skin were stomach-turning. And gods, he was wounded... they hadn't just come to kill him, they had come to torture him. Some of his clothing had been torn, clearly intended to expose more of his skin to the sun. A patchwork of wounds crisscrossed up his right arm, and the very tip of one wing had been cut—cut off? Maybe. It was hard to tell through all the blood.

"Help me," I choked. "To the house. Out of the sun."

I was only capable of assembling fractured handfuls of words at a time.

Farrow—gods bless him—did as I asked. If he was put off by being this close to a vampire, he didn't show it. Together we dragged Vale up the steps to the back door, which led into the library—the very same room he had brought me to the first time I came here. Vale was incredibly heavy, even with both of us carrying him, and I was grateful that he appeared to be at least a little bit conscious, because he seemed to be trying to help us—albeit poorly. Still, we couldn't hoist him onto one of the couches, and instead had to settle for laying him on the floor as gently as we could.

The wounds somehow looked even worse in here, but to my relief, they had stopped spreading once he was out of the sun.

But he wasn't moving. He was only barely breathing.

"Lilith..." Farrow said quietly.

I looked up. He peered out the window, to the dead bodies lying in the yard. At first, I thought maybe he was sickened by what we'd just done—we'd killed, after all—but when he glanced back at me, it held something harder than guilt.

"An acolyte," he murmured. "Vale killed an acolyte."

The reality of what had just happened hit me.

Vale, a vampire, a child of Nyaxia, had just murdered a high-ranking devotee of Vitarus.

I had already been pushing my luck with my experimentations with vampire blood. I had been so careful at first to hide my work, to make sure I didn't touch the blood long enough to attract the attention of a scorned god. And if a few vials of blood might have been enough to earn a god's wrath...

...Imagine what the death of an acolyte could do.

Cold, cold dread fell over me. Some gods were fiercely protective of their acolytes. Others ignored them. Most,

Vitarus included, fell somewhere in the middle, depending on their mood and your luck. He might not notice what had happened here. But if he did... few things were considered more insulting to a god than the murder of what they considered theirs—especially by someone touched by their greatest enemy.

My hands went numb, like all the blood had drained from my extremities.

"I don't know what to do."

I didn't mean to speak aloud. I always knew what to do. Always knew the next logical step. But right now, logic seemed so far away. There were so many problems, all so big. I couldn't find the answers.

I turned to Farrow, wide-eyed, and swallowed a stab of guilt at the sight of him.

Farrow. Poor Farrow. I had barely looked at him before. He was covered in blood, too. One arm looked injured.

But his hand fell to my shoulder, giving it an encouraging squeeze.

"You will," he said. "Just think."

Farrow did always make me want to believe him, and that counted for something.

I drew in a breath, let it out, and stood.

"We need to burn the bodies."

Maybe if we burned them fast, Vitarus would never know. Gods were fickle and flighty. They had a whole universe to pay attention to, after all. Maybe we'd gotten lucky, and this one hadn't noticed us today.

But if we weren't...

I looked down at myself. My blood-stained hands.

I'd make sure the blame would be mine. If Vale and I stayed away from town, I could pray that we would draw Vitarus's attention, miles away from Adcova.

And if we only had a little bit of time before we attracted the attention of Vitarus, then we needed to use it.

"Here." I shoved my bag into Farrow's hands. "Take this back to town. The medicine in it…"

Did it work? Did I know for sure? It worked on the mice. Gods, I hoped it worked on…

I had to blink away Mina's face, because the thought of her almost made me fall apart.

"It works," I said. "Guard it. Don't destroy it. Don't let anything happen to it."

Farrow's brow furrowed. "Are you sure?"

Sometimes, those three words coming from someone else would be an admonishment. From Farrow, it was an actual question, spoken with the understanding that he would accept whatever answer I gave him.

I wasn't sure. And I was a terrible liar. But I still replied, with as much confidence as I could muster, "Yes. I am."

For decades, this town had thrown its faith blindly into gods that had done nothing for them but curse them. Now I'd give anything to cast that faith into those little glass vials.

"Go," I said to Farrow. "Be quick. You don't have much time."

"What about him?"

Vale lay listless on the floor. Strange, how none of this—the dead bodies, the blood on my hands—terrified me as much as the sight of him in this state.

"I'll take care of him. And the bodies."

I heard all the judgment in Farrow's silence.

"No arguing," I said, before he could protest.

But it wasn't Farrow that argued.

"Go."

The voice that came from behind me sounded nothing like the deep, smooth sound that had greeted me when I first walked through these doors months ago. Still, my heart leapt to hear it.

Vale's eyes were slitted, like he had to fight to keep them open.

"Go, mouse," he rasped out.

No. The word was immediate, definitive. If there had been any shred of doubt within me, the sight of Vale, struggling to even speak, destroyed it. I would not leave him like this.

I forced a smirk. "I owe you roses," I said. "We had a deal."

The spasm of muscles around Vale's mouth could barely be called a smile.

I led Farrow to the door before either of them could argue with me more. Farrow knew he couldn't change my mind about this, either. Before he left, he reached out and took my hand. Squeezed it. I had to close my eyes. The emotion on his face made me uncomfortable.

"Thank you." My voice was strangled and choked.

"Good luck, Lilith," he said, in a tone that sounded a lot like a goodbye.

CHAPTER EIGHTEEN

When Farrow was gone, I dropped to my knees beside Vale.

"I—I don't know how to help you. Do you have medicine, or—"

"Burn them first," he wheezed.

"Not with you like this."

"*Burn. Them.*" His gaze slipped to the parted curtains—to the sky. No sign of a god's anger now, but the longer we waited, the greater chance there was it would come.

I knew what he meant: *we don't have time.*

So, reluctantly, I did as he asked. It took longer than I'd hoped. The bodies were heavy. I struggled to drag them far enough from the house to keep the flames from spreading to the building. There were many of them.

By the time I was done, the red cast from the fire doused the entire estate. It was dusk, the sky pink as scar tissue, gritty with smoke. I rushed my work and hurried back into the library the minute I was sure the fire wouldn't take the house with it. Sooty sweat plastered my shirt to my skin. I was panting. I had worked so, so fast. But when I saw Vale lying there, right where I had left him, I thought, *I made a mistake. I should have healed him first.*

Still, I breathed a sigh of relief when he turned his head laboriously to look at me.

Did he look a little better? A little?

"Medicine," I demanded. "Where?"

"Study," he said, in a thick, scratchy voice. "Third drawer."

The drawer, of course, was a mess—I could barely get it open for all the clutter. I cursed him for it as I rummaged. I didn't even know what Obitraen medicine looked like. Finally, at the bottom, I came across several glass bottles. Most held blue-white liquid that glowed faintly. When I touched them, I shivered a little, like the magic was calling to some dark part of myself.

I wasn't sure if there was a difference between them, so I filled my arms and brought all of them back, dumping them on the coffee table beside Vale.

"Which?"

Vampires did have incredible healing ability. Vale was able to move a little bit now—at least enough to select the bottle he needed. He shot one back like strong alcohol, hissing and cursing.

"Upstairs," he said.

"You shouldn't move—"

He glowered at me. "*Up. Stairs.*"

I rolled my eyes, but managed to get him into his bedchamber, though he leaned heavily on me the whole way. I helped him strip off his bloodstained clothes, conscious of every wince as coarse fabric clung to raw skin. Vale had lit the candles in the room with a wave of his hand when we walked in—the flames were strange and white, and moved a little differently than fire did. They cast silver over his bare flesh, and as I watched him withdraw another glass bottle and tend to the worst of his wounds, a knot formed in my stomach.

I'd come to admire Vale's form so much—his blood, his body. But now, the blood that I had found so breathtakingly entrancing covered the flesh I had found equally stunning in grotesque smears. A dark, taunting mimicry of everything I'd grown to find so beautiful.

He didn't want my help, at first. But he was being ridiculous—he couldn't even reach the worst of his burns. I snatched the medicine from his hands, and after a few minutes of grumbling, he let me take over dabbing the potions onto the wounds of his back and shoulders.

Honestly, I was grateful that he had the energy to argue. And maybe he was grateful that he didn't have to do much of it.

Nyaxia's magic must have been powerful, because the healing was miraculous. Still, Vale's wounds were deep, brutal. The cuts from swords were bad enough, but the sun had inflicted the worst of it. It had been a bright day today. It left seeping, blackened patches over his skin. The potion helped, closing the open patches of skin, but still leaving behind dark purple marks.

It was my fault this had happened.

This thought solidified in my mind fully formed, a single truth.

I should have been more careful. My colleagues at the university, my parents, my sister had always been right about me—my enthusiasm made me careless. I had been so excited about my discoveries—about Vale—that I hadn't hidden my work. I forgot to be afraid.

A mistake.

"I shouldn't have allowed this to happen," I said, quietly, as I worked.

"None of this was your fault, mouse. Do you think this was the first time humans came to my door blaming me for whatever tragedy they faced that decade?" He glanced back at me with a wry smile. "Humans. All the same."

I hated my own kin in this moment. But not as much as I hated myself.

I moved on to another burn, watching Vale's skin twitch and burn beneath the silver liquid.

"You should have left," he said. "I would have survived."

"No, you wouldn't have."

"Your friend wanted you to go with him. More than he expressed, I think."

I shrugged. It didn't matter what Farrow wanted me to do.

Then Vale added quietly, in a tone of voice I could not decipher, "He is in love with you."

My eyes stung.

I couldn't even deny it. And what good had it ever gotten him?

"It's just old feelings," I said. "We were together for a while. But it ended."

"Why?"

"He wanted more than I could give him."

A life I couldn't live. A heart I couldn't free. A role I couldn't play.

Vale nodded, as if this made sense to him. We didn't talk for a long time. I was working on the last of the burns when he finally spoke again.

"I decided to go back to Obitraes."

My heart stopped. My hand slipped. Just as well, because he turned around, his amber eyes cutting through me.

Why was it suddenly hard to breathe?

"Why did you change your mind?" I asked.

His fingertips ran back and forth over the back of my hand, absentmindedly. His gaze slipped away, to the strange white flames.

"Have you ever been in love?"

My brows leapt. I wasn't expecting that question. I didn't know how to answer.

I loved Farrow. He was one of my closest friends. But was I ever *in* love with him?

Strange that it wasn't Farrow's name on my lips as I watched Vale's serious profile, silhouetted by the white firelight. And I was grateful that he didn't wait for my answer — or perhaps, heard the truth in the lack of one.

"I had only one great love," he went on. "The House of Night. I helped build an empire. I shaped it with my blade and blood. I gave my king, my men, and my kingdom my unquestioning and all-consuming devotion. If you have ever loved something that much, you know that there's no wine sweeter, no drug stronger. And when it fell..."

His throat bobbed. He stared into the fire.

"I was angry for a very long time. I came here to escape the memory of my failure—but then I spent every day dreaming of returning to the House of Night. Dreaming of rebuilding what I had let fall."

"Then it's good you're going back," I said, my mouth dry.

It's good, I had to repeat to myself.

Vale needed to leave. He needed to leave to save himself and to save us. He'd murdered an acolyte of the White Pantheon. Maybe Thomassen had been right. Maybe Vale's presence here—his presence as a tainted child of Nyaxia—did only worsen our fates.

What did it say about me that, despite all of that, the thought of Vale leaving made my soul ache?

I fidgeted with the rag because I needed something to do with my hands. "You must be happy to go home."

Vale's gaze turned to me.

"I thought I would be," he said. "But perhaps they, like your friend, want something I can't give them. Maybe they want some part of me I have already given to someone else."

I let my eyes fall down to the bedspread—to my hand pressed against it, and Vale's atop it, those graceful fingers stroking the shape of the delicate bones at the back of my hand the way a musician stroked the strings of an instrument.

My heart thrummed so loudly in my chest.

And looking away didn't save me from Vale's stare, because I could feel his eyes the way one can sense a wolf stalking them in the forest.

Except I wanted to be caught.

The bed shifted as he turned to face me fully. He leaned a little closer. His scent surrounded me.

"Why did you come here," he asked, "when you realized they'd come for me?"

"Because my work isn't done."

A lie. It was as done as it was going to be.

"Look at me, Lilith."

Vale rarely said my name. The sound cut me down to the bone, shivered and swirled just as it did when he wrote it over the page.

Look at me, my sister had begged.

And I felt just as frightened now, as I forced my eyes to lift, forced myself to meet Vale's stare.

Once it had me, I was utterly ensnared. I couldn't hide.

Run, a voice inside me whispered.

Stay, another begged.

As Vale's fingertips reached for my cheek. Stroked my cheek, my jawbone. Brushed the bridge of my nose. He wore the same expression that he had the day I showed him his blood—the day I realized, for the first time, that the only thing more beautiful than his blood was the expression of amazement on his face.

Tears pricked my eyes.

"You want more than I can give you," I whispered.

"I can't imagine that ever being true," he murmured. "Because I want only you, Lilith. Whatever of you I can have. I'll take one night. One hour. One minute. Whatever you want to give me. I'll have it."

My breath was ragged, choppy. It burned in my chest with all the emotion I realized I could no longer smother.

I had never been enough.

I had never been able to give any of them enough— enough time, enough love. Everyone gave up so much trying to get more from me, and now I did the same for them. From the moment I was old enough to understand my eventual fate, I made every decision knowing this. Knowing that I couldn't be enough. Knowing that I would wither too fast, like a flower in an early frost.

I didn't realize how much I had liked that Vale didn't see that in me until this moment, when I knew that it had to end.

"I'm dying," I choked out.

I didn't know why I said it. It didn't really matter, now, when he was leaving and the gods were damning us and the whole world seemed to be ending.

"I've been sick my entire life. Every year I don't know if it's the last. I've been leaving this world since I was brought into it. No one wants to believe it, but it's the truth. It always has been. I'm—I can't *stay*."

You're asking for more than I can give.

His hands had come up to my face. He held me firmly, so I couldn't look away from him.

I could always see the moment things changed, once they knew—the moment they started grieving me while I was still alive, the moment me standing in front of them stopped being enough.

But his gaze was firm.

"Whatever you wish to give me," he repeated, slowly, like he wanted to make sure I understood. "I'll have it."

I didn't know that I had been waiting my entire life to hear those words until now.

I wasn't accustomed to goodbyes. I never thought I would need to be the one to say them. It's so much easier to be the one who leaves first.

I could leave now and spare myself a goodbye I wasn't ready for.

But instead, I put my hands on either side of Vale's face, a mirror of how he held me.

I pulled him close, and I kissed him.

CHAPTER NINETEEN

I wasn't sure why I had expected the kiss to be fierce and animalistic, but that first one was quiet, gentle. Sweet.

Vale's lips were softer than I thought they'd be. His beard tickled my chin. At first, he just brushed his mouth over mine, like he wanted to start by knowing the shape of it, knowing the way I tasted.

Then, his lips parted, the kiss deepening, the touch of his tongue — shockingly shy — meeting mine. My head was cloudy and fuzzy in a way that had nothing to do with my exhaustion.

A serrated breath ghosted over my lips — and that, that one little sign of the intensity of his desire, lit something on fire inside me. Suddenly Vale's closeness, the warmth of his bare skin, the taste of him, the smell of him, overwhelmed me.

A tiny, wordless sound escaped my throat, and I kissed him back this time. Harder. Deeper.

He met my fervor with enough enthusiasm to leave me breathless.

He held my face firmly, his tongue exploring my mouth, each kiss bleeding into the other. Gods, I had never kissed anyone like this—each movement so intuitive. I never had to stop and guess what he wanted. It was the kind of ease I thought other people must always feel.

One of his hands moved to the back of my head, tangling in my hair. The other wandered down to my waist, his thumb slipping between the buttons of my shirt, brushing my bare skin. That one touch made me gasp.

His tongue rolled against mine, then he withdrew. In my fervor, we'd both fallen back onto the bed.

Everything was hazy, distant.

"You're injured," I said softly.

His chuckle was low and thick. "Incredible how much better I already feel."

But his smile faded, and he gave me a long stare—and I knew what this wordless silence meant, the question he was asking.

I parted my thighs, opening myself to the rigid press of his desire between us.

His eyes darkened, the desire in them so sharp it cut me open, and it occurred to me that maybe I should be afraid—that maybe the hunger I was seeing in Vale's expression, feeling in the way he held me to the bed, was about more than sex.

I wasn't, though. No, the fear came from somewhere else. Not from Vale's roughness, but his tenderness.

He smoothed a strand of hair from my forehead.

"You're shaking, mouse."

I slid my fingertips beneath the waistband of his trousers, a light touch over the flesh of his abdomen—soft skin, hard muscle, trembling faintly.

"So are you."

My voice was rough, low. Vale lowered his head a little

when I spoke, like he wanted to feel the words over his lips—stopping just shy of meeting them.

Neither of us moved. Not meeting that almost kiss, not pulling away, our hands both at the buttons of each other's clothing but not unbuttoning them.

I watched Vale's face, the panes of his features outlined in blue-silver licks of light that reminded me of the outlines of the roses I gave him. Even with the wounds remaining, he reminded me of a statue—a work of living art, carved from stone, subject to none of the atrocities of time or nature. He was eternity while I was impermanence—a being that embraced the mysteries I spent my entire life stifled by.

How could a being look so similar to a human and yet so stunningly different?

And yet...

Yet...

The corner of his mouth tightened. It should have been a smile, but the expression was so sad it gutted me.

"I always wondered what you were thinking," he murmured. "When you look at me like that."

"Like what?"

"Like I'm a formula to be solved, and you're very intrigued about the answer."

At this, I couldn't help but smile. "Intriguing is the word."

A wrinkle formed between his brows. "An acceptable one?"

The question struck me hard—struck me because I wasn't prepared for it, for him to ask it that way, shy and tentative.

Like the answer meant something to him. Like the answer meant everything to him.

"Yes," I said. "It's remarkable."

I could never solve Vale and his many mysteries, but I loved them all the same. And in these complexities, I saw a mirror held up to all the things that did not make sense within myself.

For the first time, I saw beauty in all the things I did not understand. And I knew that Vale saw beauty in all those things within me, too.

I slipped my palm up his abdomen and relished the way his muscles twitched beneath my touch.

"I'd like you to kiss me again," I said. "And I'd like these clothes off."

"Hmm." He hummed feigned reluctance against my mouth, but only for a moment, because it was quickly swallowed by his next kiss—and this one was brutal, hard, demanding. He kissed me like we didn't have any time. Like he was mortal.

His hand slid up my shirt, large palm flattening over my stomach, as if not sure whether he wanted to go up or down— gods, I wasn't sure where I wanted him to be, either. I wanted both, and quickly.

I broke away from his kiss just long enough to tear my shirt off over my head. A rush of cold air hit me as Vale pulled slightly away, despite my rough exhale of protest.

Those amber eyes moved over my body, taking in my bare skin. The hunger in them was unmistakable now.

You should be afraid, a voice whispered in the back of my mind.

But I'd never been afraid of death.

No, that hunger just fueled my own. My breasts were peaked with desire, begging to be touched—my core slick, begging to be filled. His eyes drank me in for what felt like an age and an instant.

Then, like his own desire had overwhelmed him, he reached for the buttons of my trousers and yanked them open.

I lifted my hips to help them off, and I had barely kicked off the fabric before his hand was between my legs.

Pleasure sparked up my spine. A whimper escaped me. My fingernails dug into his back, and my body, of its own accord, jerked to be closer to his—even though he held me down to the bed.

He let out a low groan.

"I was right that night." His mouth curled into a smirk against mine. He sounded satisfied with himself. "You do want this."

He'd been right then, and I wasn't even ashamed of it. I'd wanted him for a long time. That night had been the first time I thought of him when I found my own pleasure in bed, imagining my own hands to be his, and it had only gotten more frequent since then.

Now that his hands were there, circling in on the epicenter of my desire... gods, it was better than I'd imagined.

It was hard to speak, hard to think. He teased me like he knew it, too, though I could see that he was also having a hard time focusing on anything but that, see it in the way his eyes hooded when my breath hitched.

"You've been thinking about it for a long time, too." My hand slipped farther beneath the waistband of his trousers, slid over smooth skin and coarse hair and settling at the rigid length of his cock straining against the fabric, responding immediately to my touch.

I took a moment—just a moment—to run my palm up and down that beautiful length, just softly enough that I knew it would be a little torturous. Just to make sure he knew we were on equal ground.

He smiled into my kiss like he knew it, too.

But then I kissed him hard and ripped the buttons of his trousers open.

Time.

We didn't have time.

And that realization seemed to crash over him at the same time it did me, because he pressed me to the bed, our kisses frantic and messy, his tongue exploring my mouth, fingers sliding into me — the sudden press of them coaxing a choked moan from my throat, my thighs opening wider for him, though he held me still when I tried to chase the friction my body wanted.

His kisses moved from my lips to my cheek, pausing at my ear — his breath rough against the sensitive skin there, teeth catching my earlobe, something I never thought could feel as good as it did in this moment — then moving down, to my throat. He paused there, tongue pressing against my flesh.

His breath was ragged. My heart pounded. I was sure he could feel my pulse there. Smell it.

For a moment, I thought maybe he might do it.

For a moment, I thought maybe I wanted him to.

But he just skimmed his mouth up my neck, my jaw, moving back to my mouth and kissing me hard. His thumb pressed down at the core of my need, and that, combined with the penetration of his fingers, sent a wave of pleasure through me that left me gasping.

Time.

I pushed him off of me, my eyes meeting his in a way that communicated all of my demands, and started to roll over onto my hands and knees. I wanted him as deep as I could have him.

But he stopped me.

"No," he said. "I want to watch your face."

I hesitated, and my uncertainty must have shown in my expression, because Vale smiled — smiled fully, to reveal those deadly fangs.

"All this time you've gotten to study me. That isn't fair."

And strange how having him at my throat, teeth one heart-beat from my blood, didn't frighten me, but the idea of letting him do that—the idea of looking into his eyes when I was so exposed—gave me pause.

But his fingers circled my bud, and I let out a strangled moan, and he smirked in a way that said he knew he had me.

And he was right. He could have me however he wanted.

I let him push me to the bed. My thighs opened around his hips. He kissed me languidly as I reached down to position him at my entrance—even the first pressure of his cock there making us both groan.

He yanked my hand back, pressing down on my forearms to hold me beneath him, and thrust into me.

I was so wet, so ready. It took just one single thrust. He was bigger than anything I'd had before, and that first thrust almost—*almost*—hurt, in every wonderful way.

I didn't even realize I'd made a sound until he reacted to it, a hiss of pleasure as he buried his face against my hair. He worked at me slowly for those first couple of strokes, my hips rolling and pressing against his movements. Forcing him deep, gasping at every new angle that he hit inside of me.

He pushed himself up enough to look at me, and my impulse was to turn my head, to look away. But he grabbed my chin, held it—held it, so that he was looking right into my eyes.

He withdrew slowly, then pushed back into me, deeper, until my hips were lifted off the bed with the force of it. Sparks shot up my spine, pleasure spreading through my core. My one free hand reached for something, anything, to hold onto, finding his shoulder and clutching so hard that surely I was leaving marks on him.

He held that pressure for a few agonizing, incredible

seconds, watching me as each minuscule shift made my breath quicken.

"Yes?" he said, softly.

"Yes," I answered.

Gods, yes. Yes, yes, yes.

He withdrew again, painfully slowly.

His next stroke was harder still. My moan came out ragged, ripped from me without my permission.

Another stroke. Faster. Forceful.

He was still watching me, his face serious and focused, and I wanted to look away, wanted to hide myself, but I couldn't—his eyes, the amber gold of a wolf in the woods, transfixed me.

Again.

He was slowly increasing his speed, his pressure. His free hand, the one that was not holding my forearm to the bed, traced the curve of my hip, my waist, circling the peaked hardness of my nipple just as he pushed into me again.

This time, my moan became a cry.

"Yes?" he asked.

"Yes," I gasped.

Still, we didn't look away from each other.

He was decoding me, solving me, the way I had solved him. I was being projected onto the wall like I had projected his blood, and I knew with a strange, terrible kind of certainty in this moment that he found me just as remarkable.

He wasn't the only one. Because even though he had let go of my chin, I didn't look away from him, either.

No, I barely blinked as he continued fucking me, every carefully measured stroke loosening in control. He was a quick study. He learned fast what I liked, what angles made my moans loudest. Learned what to give me when desperate, nonsensical pleas tumbled from my lips, even when I myself didn't know.

Every muscle of my body, every shred of awareness, rearranged around him. The pleasure was unbearable, agonizing. I wanted to throw my head back and scream his name — I wanted to bury my face against the smooth expanse of his skin and breathe him.

I didn't. Because I couldn't look away from him, watching him watching me, memorizing each other.

And gods, he was beautiful. More beautiful than his blood. More beautiful than his admiration. All of it was dwarfed by the way he looked slowly unraveling, losing himself in his pleasure the way I lost myself in mine, tethered only to each other.

I clutched his shoulder now, and his fingers were tight enough around my arm to leave marks on me. My legs folded around his hips, urging him into me faster, harder. The headboard banged against the wall, an increasing rhythm that echoed my heartbeat.

His lips found my cheek, my throat, my mouth, stifling my cries. And yet he pulled away again, right as he rushed to that pinnacle, his cock driving into me so hard that he had to clutch my waist to keep from sending me against the headboard.

He met my eyes. And I knew he wanted to see the conclusion of this experiment — as much as I did.

"Yes?"

His voice was strained, like it took a lot of concentration to form even that small word.

I took his next stroke with equal force, pushing against him, contracting around him.

"Yes," I choked. "Yes."

And he pinned my shoulders down as I lifted my hips to

receive those final thrusts, and we watched each other's faces as we came together. I had to fight to keep my eyes open through the explosion of pleasure that left sparks of white over my vision, that tore a cry from my throat that must have echoed down the ancient empty hallways of this house.

But gods, it was worth it to make sure I saw him, eyes both distant and sharp with ecstasy, looking as if he had seen his goddess herself.

He pushed deep as he came, and I wrung myself around him as if to make sure I gave and took every last shred of pleasure.

The world went quiet. Reality came back in blurry pieces.

Vale's head dropped, his forehead pressing against mine. His muscles trembled a bit, which I noticed with a pang of guilt. He'd strained himself more than he should have so soon after his injuries, magical potions or no.

He rolled off me and, as if it was nothing other than instinct, his arms folded around me, pulling me onto his chest.

I had never liked being held much. I found it too hot and restrictive. But Vale's body was just the right amount of warm and cool, just the right balance of soft and firm. It felt like it was built to accommodate the shape of my own.

I let him hold me, and as my eyelashes fluttered with a sudden wave of exhaustion, a terrible dread settled over me.

Vale had been my experiment, my question to be answered. I thought it would be easier to let go of him if I could understand his every unknown. But he was a question that had no answer. And every answer.

Vale wasn't a cure for anything. He was a whole new disease, one I'd carry with me to my inevitable end.

I didn't want to let him go. I didn't like goodbyes. Easier to be the first one to go.

But they come for us all, anyway.

CHAPTER TWENTY

I really didn't mean to sleep.

I didn't have time for it. I never did—maybe that was why my body forced it upon me. One moment, I was allowing Vale to hold me. The next, I was blinking blearily into the shadows of his bedchamber. I hadn't spent much time in this room. It was just as cluttered as all the others—full of books and weapons and mismatched artifacts, like he'd just run out of space to put the vast quantity of things he'd collected over his long life and just shoved them wherever he could.

The smile came without my permission.

Vale. Someone who collected knowledge just like I did. I felt like a failure of a scientist for not realizing what I was seeing the first time I came to this house. I thought it was just full of clutter. But no, all these things had touched him in some way. He was careful about what he kept.

He slept now.

I knew that before I even looked at him. I could feel the

steady rise and fall of his chest beneath my head. It was a deep sleep. Good. He needed it.

I didn't want this moment to pass.

I blinked away sleep and stared into the room. The blue light of white flames now flickered alongside a warmer accent. My eyes fell to the windows. Dim light seeped beneath the curtains. Daylight from an overcast sky.

Daylight.

"Shit," I hissed.

How? How could I have slept so long?

When I pushed myself up, a wave of dizziness greeted me. My whole body protested. The hard realities of our situation crushed me one after the other.

The dead priests I had burned.

The medicine.

Vitarus.

Time. We didn't have *time.*

And I had let myself fall asleep.

Shame flooded through me. Embarrassment, that I'd let myself be distracted for so long—that I'd let Vale see—

I stood abruptly, ignoring my shaking knees and the sway to my step as I crossed the room.

I heard rustling fabric as Vale stirred behind me.

"Where are you going, mouse?"

His voice was weak and sleep-slurred, and I heard those things before I heard the joking lilt to his tone. He was still injured.

"I slept too long."

He laughed. "I already know you well enough to know that is never true."

It was true right now, when the world was falling apart. I went to the curtains and peered through them, careful not to let sunlight fall over Vale's bed.

The window overlooked the back of Vale's estate grounds. The charred remains of the bodies I had burned were a smear of ashy black bones.

I raised my gaze, and my throat closed.

No.

My knuckles trembled around the handful of velvet curtain.

Vale said, after a moment, "What is it?"

I didn't even know how to answer him.

The end. That's what it is.

I had seen once before what the sky looked like before a god appeared. I knew in that moment, all those years ago, that I would never forget the sight. And I knew it now, too, that I would never forget this one.

It wasn't overcast, like I'd thought. The light had seemed strange because the sky was warped. Sunlight hit the ground in mottled, jerking flecks. Clouds circled in unnatural swirls in the distance, drawing tighter and tighter, and though the thickening mist at its center seemed like it should be dark, like storm clouds, instead it cradled distant fragments of bright yellow light—like little shards of lightning, floating suspended in the air, moving in slow ripples rather than jagged cracks.

The center of it was not over this estate.

No, it was miles away. One look, and I knew it hovered over the town of Adcova.

I couldn't move. Panic settled deep in my bones.

"Lilith?"

Vale rose and approached me. I felt his warmth behind me, even though I couldn't turn to look at him. He peered through the curtain, staying away from the light, and released a long exhale.

"I had hoped..." he murmured, and then let himself trail

off. Because we had both hoped the same thing—that Vitarus had long ago decided he didn't give any care to Adcova, and he'd continue to ignore us. Any encounter with the gods was a gambling game, and we had lost.

Of course he didn't listen to decades of prayer and pleas for mercy. Of course he didn't listen to dozens or hundreds or thousands of sacrifices in his honor.

This. *This* is the thing he would notice. What a cruel, ridiculous joke.

Our sins had not escaped Vitarus, and they would not go unpunished.

I closed my eyes for a long moment.

"You should leave now."

My voice sounded strange when I said it. The words hurt more than they should have.

"I'll go with you," Vale said. "Help you."

"You can't help. It would make everything worse."

"And what do you plan to do?"

My lips parted, but I tripped over words I didn't have. What did I plan to do? What *could* I do?

"I don't know when you were planning on leaving, but make it now. Right now."

"Lilith."

He didn't say, "Look at me," but I heard him ask for that in his tone of voice. And despite my better judgment, I turned.

Vale looked... sad.

I expected frustration. The same expression I was used to seeing on the faces of the people unfortunate enough to love me. But Vale... he just looked resigned, like he knew why I was doing this and that he couldn't stop me.

"I need you to know—"

"I don't have time for this."

"Listen." His hand fell to my arm—holding me gently. Did he know that it was the same place he held me down last night? "I know you, Lilith. I know that no one can make this decision but you. But let me give you all the information to make it with."

I should have stopped him, but I didn't.

"You could leave with me," he said.

I knew he was going to say it. But it still ached to hear.

"If we run now," he went on, "you will be gone by the time Vitarus shows his face. We could draw him away."

I swallowed thickly. "To Obitraes?"

"Anywhere. Everywhere. It doesn't matter. None of the gods of the White Pantheon can touch Obitraes. But if you wanted to go somewhere else, we could do that, too."

There was nowhere one could hide from a god.

And it was foolish and naive to think that Vitarus wouldn't destroy my home, a town that had already earned his ire, out of nothing more than petty boredom, whether I was there or not.

Vale knew this just as well as I did.

"You aren't a stupid man, Vale," I said quietly, and he winced.

"No," he said. "Just a desperate one."

He stepped closer, his body now flush to mine. His hand released my arm and moved to my chin—touched it more gently than he had last night, but the grip seemed just as inescapable as he looked into my face, our noses brushing.

"You do not have to do any of this alone," he said.

It wasn't the first time someone had said that to me. But it was the first time I really wanted—needed—to hear it.

"I don't want you there," I said. "It would be dangerous. You're one of Nyaxia's children. Any god of the White Pantheon would hate you for it, including Vitarus. The best thing you can do for me is get far away from here and never come back."

My words were sharp and clipped and cold. The same voice I would use when I told Mina I could not stay with her or sent away Farrow when he asked too many searching questions. Hard as iron.

That tone would usually send them away with a scoff and a shake of the head.

But Vale didn't let go of me.

"It must be hard," he murmured. "To bear the weight of so much affection in a life so short."

My eyes burned ferociously. I had to squeeze them shut, had to clamp down on my sudden shuddering inhale.

No one had ever seen that before. The love in my cold absence. And it was always so easy to just let them believe I didn't feel it.

All this time I thought I had been studying Vale, but he had been studying me.

For one horrible moment, I clearly saw exactly how precious this... this thing we had built was.

I would never meet anyone like Vale ever again.

Stay, I wanted to tell him. *Stay with me. I don't care if it damns us both. I don't care if it damns my entire town. Stay, stay, stay.*

But I pulled away from him and went to my pack, which was now discarded at the foot of the bed. The rose was a little crumpled, the petals squished to one side. I owed him two. I only had one today, this ugly thing, lopsided and deformed, but still—always—living.

I hated these roses. I hated them so much.

Vale reached for me, but I only pressed the rose into his hand.

I met his amber eyes.

Stay, my heart begged.

"Go," I said. "I'm leaving, and you should, too."

Vale knew me better than Farrow. Better than Mina.

To his credit, he did not ask me not to go.

YOU CAN FEEL it in the air, when a god is near. It breaks and shivers, like invisible lightning hanging in your breath, cracking over your skin.

It felt exactly the same as it did that day all those years ago.

I rode as fast as my poor exhausted horse could carry me. The beast was near collapse by the time I arrived back in Adcova, already nearing sunset. I practically flung myself off him when we reached my cottage, throwing open the front door, calling frantically for Mina.

I checked my study, her bedroom, the kitchen. The house was empty.

I wanted to believe she just went to town. But the hairs on my arms stood straight upright.

Maybe a part of me knew what I would see when I opened the back door, the one that led to the fields.

The door opened, and for a moment I was a child again, standing in this doorway, watching my father on his hands and knees in those wretched fields, feeling this same horrible sensation of divine dread.

Mina was out there in that exact same spot, her back to me, surrounded by wild rosebushes.

The air was still. Silent.

She held herself upright for the first time in months. There was no dusting of ivory skin in the dirt around her.

"Mina," I called out.

My voice wavered. My steps did, too, as I approached.

Mina didn't turn. Her head was tilted up.

Above us, the clouds circled, circled.

And there, at their center, was Vitarus.

PART FIVE

THE FIFTH ROSE

CHAPTER TWENTY-ONE

Vitarus was beautiful.

All gods were beautiful, and all mortals knew this in theory. But when people say the gods are beautiful, you imagine it as the beauty of a human—perhaps even the beauty of a vampire, ageless and perfect.

No. No, that wasn't right at all. Vitarus's beauty was that of a mountain range or lightning storm, the beauty of the sun reflecting off the horizon of a rolling plain, the beauty of a fierce summer storm that kills half the town's livestock, the tragic beauty of a stag's body rotting and returning to the earth.

Vitarus was beautiful the way death was surely beautiful moments before it took you.

He lowered himself to the ground, though his feet didn't quite touch, hovering just above the tips of the sparse grass. He was tall and foreboding. His hair and eyes were the ever-shifting gold of sunshine and wheat fields, his skin gleaming bronze. He wore loose trousers of silk and a long, sleeveless robe that looked as if it could be either green or gold with every blink, which he left open, exposing a lean torso covered with the silhouettes of flowers and leaves. His hands and fore-arms were darker than the rest of him, all the way up to the elbow—they looked different from each other, though I couldn't place why, not when I was so preoccupied with my own overwhelming fear.

A shimmering white mist surrounded him. Water vapor, I realized, when he ventured closer and the damp of it clung to my skin. The grass rustled, greened, withered beneath his feet.

For a moment, the presence of him paralyzed me.

Then his world-shattering gaze, disinterested and cruel, fell to my sister. Mina cowered like a deer cornered by a wolf, and that sight awoke every wild protective instinct in me.

I didn't even remember running to the field.

"Go," I bit out, shoving Mina aside as I fell to my knees before Vitarus. "Go, Mina."

I didn't look back long enough to see if she'd listened—where was there to run, anyway?

And I couldn't, even if I'd wanted to, when Vitarus's eyes locked to mine. They were a million colors of the sky and earth, every shade of radiant sun and coarse dirt.

"It wasn't her," I said. "She's innocent. She did nothing to offend you."

His gaze was so entrancing that it took me too long to remember to bow my head. I lowered my chin, but a firm grip tilted it back up. Vitarus's skin against mine ripped a gasp from my throat.

A dizzy spell passed over me—a wave of fever, weakness. Death's breath over my skin, a too-familiar sensation I hadn't felt quite this strongly in a very long time. My eyes fell to the darkened skin of Vitarus's forearm, and the nature of his hands, the thing I hadn't been able to place moments ago, hit me: this hand was decay, his skin mottled and purpling, crawling with insects. The other was dark with the rich hue of soil, roots winding up his muscled forearms like veins, hints of green sprouting at his fingertips.

Decay and abundance. Plague and vitality.

He held my chin tight, not allowing me to look away.

And then, after a long moment, he smiled.

"I remember you. My, how easy it is to forget how time moves for you. Fifteen years. A blink, and yet an age. How quickly you grow and wither."

His thumb stroked my cheek, and the flush of fever flared. My lashes fluttered, and for a moment I saw my father kneeling in these very fields, just as I kneeled now.

"You were just a pitiful ailing lamb then. Death walking in a little girl," Vitarus crooned. "And now, look at you. Time is so kind to humans. And so cruel."

He released his hold on me, the fever falling from the surface of my skin. I let out a ragged breath.

"No one offended you here," I said.

Vitarus's smile withered.

"One of my acolytes has been slaughtered. And you... you both stink of my traitorous cousin Nyaxia's stench." His eyes lifted beyond me—to the skyline of Adcova. "This whole town reeks of it."

"They had nothing to do with any of it," I choked. "They've suffered enough. Please."

I couldn't think of what to do, so I'd beg.

It was the wrong thing.

"Enough?" Vitarus said, incredulous. "*Enough?* What is it to suffer *enough?* The mouse suffers at the fangs of a snake. The snake suffers at the claws of a badger. The badger suffers at the teeth of a wolf. The wolf suffers at the spear of a hunter. There is no such thing as *enough* suffering."

His words were vicious, and yet his tone, somehow, was

not. He seemed genuinely perplexed by my statement, as if the idea that suffering could be cruel was foreign to him.

A hysterical wave of sympathy passed over me—because he, like me, struggled to understand human nature. Maybe we were both so bad at it that my entire town would perish because of it.

"Is that all we are to you?" I said. "Animals? Would you waste the lives of animals the way you have wasted the lives of the people you have killed here?"

Vitarus's face went cold.

"You speak to me of waste," he sneered. "The blood of one of my acolytes has spilled here. You stink of the bitch who betrayed me. I have fed your people for millennia. Sheltered you. Given you purpose. And yet you spurn me. Disrespect me." He looked around, lip curled in disgust. "I never understood the others' fondness for your kind. What would spring from this soil if this miserable assemblage of stone and wood wasn't here? Perhaps I should prefer to see that." He let out a low laugh. It sounded like the wind through the trees. "That is the mistake of my kin. Assuming that humans are more interesting than any of the other millions of forms of life in this world. No. You are no more interesting. Simply more trouble."

His gaze fell back to me, and whatever he saw in my face made him laugh again, mockingly.

"You should see your face, little girl. Such hatred." He plucked one of the roses from one of the bushes and twirled it between his fingers. The petals rustled and flourished, multiplying until they fell gently to the soil, the vine of the stem wrapping around his arm. "A flower doesn't hate. It fulfills its function, and then returns to the earth without a fuss."

I did hate him. I wanted to spit in his face and curse at him and strike him. If only killing a god was as easy as killing his acolyte.

But the thought of Mina flashed through my mind. Farrow, and the wild risks he had taken for me. My people, and the illness that would devour them all. And then the thought of Vale, and I prayed that he was far away by now from the grasp of gods that resented him.

I hated Vitarus. But what I felt for them was stronger than my hatred.

No, I couldn't kill a god. I couldn't appease him with empty apologies. I certainly couldn't move his heart to compassion.

But...

"I'll make a deal with you," I blurted out.

Vitarus paused, his interest piqued.

Gods weren't compassionate or logical. But they were bored. They liked games, liked bargains.

I didn't let my hope show as his head tilted, a slow smile spreading over his lips.

"Ah, just like your father," he said. "You know, he made a deal with me a long time ago, too."

CHAPTER TWENTY-TWO

A *deal?*

My mind grabbed onto those words and didn't let go.

A deal.

Not a punishment. An exchange.

It seemed like such a small distinction, and yet it reframed everything I knew of what had happened between my father and Vitarus that day. The story I had told myself for fifteen years—that my father had cursed his god, and by terrible chance, that god had decided to curse him right back—was false.

My father had made a choice.

Betrayal skewered me.

"A deal." The word was scratchy in my throat. "He made a deal with you."

Vitarus's eyes glinted with interest, peeking through his boredom like sun through the clouds. "You did not know?"

I said nothing, but I didn't need to speak for a god to know my answer.

He laughed, the sound rain on the fields. "You came here hating me for my cruelty. But how your heart changes when you realize it was your own father who damned your people."

He couldn't have. He wouldn't have. He...

But my fingers closed around the branches of the rose bushes, thorns coaxing blood to my fingertips.

My strange roses, that grew right here in the spot where Vitarus had stood, all those years ago. I had thought that they grew here because a god had once stood upon this soil.

But...

My father had been so upset by the crops he couldn't save. The fields he couldn't fill.

Vitarus saw the realization in me. In this moment, the only creature I hated more than my father was him—for the utter delight on his face.

"All the things he was willing to give up for some fertile soil," he crooned. "I told him that life requires death. He did not care."

Vitarus twirled the rose between his fingertips. The vine now wove all the way down his arm, the blossom and leaves so swollen they filled his palm.

"Beautiful, aren't they? Shame they aren't edible. Tell me, little girl, was he disappointed by that?"

My eyes burned. My stupid, selfish father. The truth was, he never even lived to see the roses. He was the first to die of the illness, and the first sprouts of these bushes poked from the earth after his death. I remembered vividly staring down at them as I walked home from his funeral, staring down at those little beads of green like they were an equation that didn't make sense.

Well, they didn't. They never had.

I crushed the rose in my clenched palm. It left smears of black and red against my skin.

All of it for nothing.

I had fought. I had studied. I had sacrificed whatever life I had left—and I had *succeeded*, I had succeeded in creating a cure, and it would be for nothing.

Vitarus tilted my chin up, his rose-covered hand sweeping the tear from my cheek, a thorn leaving a salt-stung scratch of red.

"Why are you so surprised?" he murmured—a genuine question. "Do you not know the nature of humans by now?"

He cradled my face like a lover, one hand on each cheek— one touch of death, one of life. I could feel both roiling inside me, surging at his touch—illness and vitality, decay and growth. My reflection stared back at me from his curious eyes, shrouded in the gold glint of his desire.

He wanted to consume me the same way he consumed withering crops. And I wanted to give up and let him.

But then, something moved over his shoulder, something barely visible within the thick cloud cover. A little glint of silver-white.

Wings.

Vale.

My stomach dropped.

Vale couldn't be here. Vitarus wouldn't tolerate a vampire in his presence. There was nothing the gods of the White Pantheon hated more than reminders of Nyaxia's betrayal.

Maybe Vale knew that.

Vitarus's brow furrowed, noticing my distraction. He started to turn, but in a fit of desperation, I turned his face back to me. His skin was violently hot, and I drew in a sharp breath to resist the urge to pull my hand away.

"I told you I want a deal," I said. "I want to terminate my father's bargain."

I couldn't offer Vitarus goods or riches. But in an immortal life, one thing becomes more valuable than all else. I heard the answer as Vale had said it to me, months ago:

Curiosity, mouse. Curiosity.

"It will be a game," I said. "If I can give you back everything that you gave my father, you will take back the plague. You'll treat our town just as you did before."

For a moment, I thought I'd miscalculated, and Vitarus's petty anger would still win. But...

There. There it was. A glint of curiosity in his eyes. Cruel amusement. His knuckles stroked my cheek—decay blossoming over my skin.

"You do not know what you are offering me, child."

"Do we have a deal?" I said.

In the distance, Vale flew faster. I could make out his shape now, hurtling through the air at impossible speed.

Vitarus could not resist it. He smiled and leaned close to my ear. "Deal," he whispered, then straightened. The full height of him, now that he stood again, nearly paralyzed me with fear. But he outstretched his hands, waiting, expectant.

My father made a deal out of desperation.

I dug a handful of the earth from the ground, then pressed it into Vitarus's waiting hands. "Soil," I said.

Vitarus's palm remained open, expectant.

My father made a deal because he was surrounded by a withering world—soil that would not give life, crops that would not grow.

I yanked a flower from the rosebushes, placing it atop the dirt in Vitarus's palms. "Flowers."

A slow, terrible smile spread over his mouth.

Vale was nearly here. I could see his face, now, desperate —his hand, outstretched, reaching for me even though he was still far away. Within it was a single flower, just a tiny dot of red and black in the distance.

"What else?" Vitarus prodded.

My father made a deal because he was surrounded by a withering world.

Soil that would not produce.

Crops that would not grow.

And a daughter that would die.

My father hated the gods for taking his livelihood. And he loved his family too much to let them go. That day, he had kneeled in the fields and looked back at me like hope destroyed, the same way he'd looked at those dead plants.

It now seemed so, so obvious.

I thought I wouldn't live to see seventeen, twenty, twenty-five. But here I was, thirty years old with a heart still beating, death matching my pace without overtaking it. Still living, just like the cursed, blessed flowers my father had left behind.

I felt like a fool for not realizing it sooner. That my longer-than-expected life was so much more than luck. When the town withered, and I lived. Why hadn't it even occurred to me to question it?

I placed my hand in Vitarus's, laid on top of the flower and the dirt.

Vale hurtled to the ground, a rough, stumbling land, just behind Vitarus.

But I had the god's attention now.

"And?" Vitarus breathed.

"Me," I said. "I give you me."

Vitarus leaned close, his lips so close they brushed mine.

"Humans," he purred. "For all your faults, maybe you aren't so boring, after all."

His kiss was fierce and thorough, his tongue parting my lips, claiming, searching. I couldn't breathe. The world dissolved. Life and death collided. He breathed into me, and his breath was growth and sun and water and light—and then he drew in a deep inhale, peeling all those things away, and coaxing forth like a fire the illness that had followed me since the day I was born. My strength withered. My lungs shriveled. My skin grew hot with fever and cold with shivers. My heart beat, beat, beat, pulsing only thin, impotent blood.

Fifteen years of illness that my father's deal had staved off now crashed back into my ailing body, all at once. Fifteen years of weakness rushing through my veins, stealing my unfairly prolonged life with it.

In the distance, I heard a familiar voice call my name.

But that shout of desperation fell far into the background as Vitarus, a lifetime later, broke our kiss.

"You have your deal, little ailing lamb," he whispered, licking my health from his lips.

And then he was gone, and I fell backwards into the newly barren soil, right back into death's embrace.

CHAPTER TWENTY-THREE

The first time I met death, I saw its face before I even
saw the face of the midwife, my mother, my father.
My death defined my entire life. It was my beginning and my only end. For fifteen years, death had hovered
with its grip near my throat—so close and yet never able to
touch me.

Well, it had me now, and its grip was fierce. It choked the
life from me in a single ruthless sweep.

Death's home was a field of blackened flowers.

You have been on your way here, it whispered, *for a long, long
time.*

"Lilith. Lilith."

Someone was calling my name. A familiar voice—a face I
wanted to turn to see. I blinked. It was hard. I saw a clearing
sky.

I blinked again.

A face.

Two faces.

Mina, my sister, her eyes brighter than they had been in a long time. Her tears were warm against my cheek.

I opened my mouth, a sudden wave of words rising in my throat—a lifetime of words I had never known how to say to her, a lifetime of affection I didn't know how to offer her. But I couldn't speak, my breath wet and burning, producing only iron-sweet bubbles at my lips.

I blinked.

Death's home was a field of flowers. A destination I had come to terms with—a path I was fifteen years late in traveling. Death walked beside me.

You seem sad to go, it said.

I stopped walking.

It was right, I realized. I *was* sad to go.

In another world, a gentle touch turned my face.

My eyes opened with great, impossible effort.

Vale leaned over me. His hand gripped mine so tight I could feel it in the next world. Maybe that made sense. Vale, like me, straddled both life and death.

And the scorched rose grasped between us was withering now, just like me.

Vale's eyes said, *Stay,* and for the first time in my life, I wanted to. I wanted to stay so badly I would die for it.

I tried to speak, tried to tell him—

He murmured, "Do you want to live, Lilith?"

I blinked and almost couldn't open my eyes again.

Death's home was a field of flowers, and someone was pulling my hand, but I wouldn't go—

Vale's voice, again, more frantic: "Lilith, do you want this?"

And I knew what he was offering me. I knew that he would accept whatever answer I gave him.

Death stopped. Turned back to me.

For a moment, I stood between them both. Vale, and death.

Do you want this?

I forced my eyes open.

"Quickly, Lilith." Vale's voice was urgent, rough with almost-tears. "Do you want this?"

I wanted life.

I wanted time.

"Yes," I choked, as death grabbed my hand.

I felt a sharp pain at my throat as the scent of dead roses overwhelmed me.

Somewhere in a world far away, my body writhed, lungs fighting for another gulp of air.

I balked. Death tightened its hold on me.

You have waited for this for so long, it told me, frustrated.

Something hot filled my mouth, pooled in my throat. Sweet, with a bitter bite.

I choked on it, sputtered.

"Drink," a familiar voice commanded — begged.

Muscles that I barely controlled swallowed. Death tasted like rose petals. It dribbled down my chin, pooling in dusty earth.

Death's empty eyes stared at me, its hand clutching mine.

I want to stay, I said.

You can't.

I need to stay.

I yanked my hand away from death's grip. Turned away from the field of flowers.

And I drew in a great gasp of air.

Vale held me tight to his chest, cradled in his arms, forehead to mine. There were tears in his eyes and blood on his lips.

"I want to stay," I choked out.

"I know," he whispered, as his mouth lowered to mine, and I faded away there in his arms, surrounded by withering roses.

PART SIX

THE SIXTH ROSE

CHAPTER TWENTY-FOUR

A million dreams consumed me. Dreams of my mother, my father. Dreams of Mina. Dreams of dusty skin on rickety floorboards. Dreams of amber eyes and silver wings.

I dreamed of mundane days and torrid nights. I dreamed of a body pressing me to a silk-sheeted bed. I dreamed of needles and vials and flower petals on a wall.

In my dreams I couldn't breathe, and I'd struggle and struggle, and then I'd put my head between my knees and choke up blood and roses.

Time passed. So much time. Flashes of the past and future, this world and the next, life and death. Pain, fever. Consciousness, unconsciousness.

I'm dead, I thought. *I'm dead. This is death.*

Or is it life?

Maybe, a voice said, *it's something in between, mouse.*

CHAPTER TWENTY-FIVE

I startled awake, choking and sputtering.

I couldn't orient myself. My entire body felt strange, foreign. My heartbeat was too loud, scents too strong, light too bright. My head pounded. My own senses overwhelmed me, blocking out all else.

Until I became aware of a hand holding mine, tightly, as if to lead me back to the world.

"Careful." Vale's voice was steady, solid. *Real.* "Careful, mouse."

Words spilled out of me without my permission. "I'm dead," I gasped. "I died. I died, and Vitarus, and my father, and—and—"

"Slow." It was only when he put his hands on my shoulders and started to push me back to the bed that I realized I had been leaning over it, precariously close to throwing myself to the floor.

I let him place me back against the headboard and a truly obscene number of pillows, though my hands were clasped tight in my lap. He eyed me with that analytical stare.

I felt *awful*. My head was spinning, I was hot and feverish, my stomach churned. My mouth was sandpaper dry, my throat raw. And my whole body... my body didn't feel the way it always had, like I'd just been put in a version of my childhood home where every measurement had been adjusted by a few inches.

But I was certainly alive.

"You remember this time?" Vale said, quietly. He wiped sweat from my forehead.

Was it the first time I had woken up?

"I..."

My head hurt so much. I squeezed my eyes shut and tried to assemble the pieces of what had happened. Vitarus. The rose bushes. The deal.

And...

Do you want to live?

The choice. The choice Vale had offered me, and the one I had taken.

"I remember."

The words were gritty because my mouth was so, so dry. As if he knew that, Vale pressed a cup into my hands. I drank without even looking at it.

It wasn't what I was expecting—water. No, it was thick and sweet and bitter and rich, and—and—

Gods, it was amazing.

I tilted my head back, practically drowning in my own frenzied gulps, until Vale gently pulled the cup away.

"Enough for now. Not too fast."

He kept his hand on my wrist, as if to keep me from drinking again. I blinked down at the cup and wiped the liquid away from my mouth. I'd gotten it everywhere.

Red. Very, very dark red. Practically black.

I recognized it right away. By sight, and... even the taste.

"It's not human," he said, misreading my expression.

"It's yours."

I'd spent months obsessed with Vale's blood. I'd know it anywhere.

"Yes," he said.

I tried to raise the cup again, and he said, "Slowly," before allowing me another sip.

I still felt horrible, but the blood helped. I took in the room around me for the first time. Unfamiliar—somewhere far from home, judging by the decor. Simple. It was a small room, and sparse, with only a few pieces of simple furniture. The curtains, thick brocade fabric, were drawn. No light seeped beneath them—it was night.

"Where are we?"

"The coast of Pikov."

My brows rose. We *were* far from home. Far from Adcova —far from the continent of Dhera, too.

I didn't know how I knew that significant time had passed. It was like I could smell it in the air—summer, the damp humidity of the sky outside, the salt on the skin of those beyond this building. I could... feel, *sense*, so much more now.

"How long—"

"Weeks."

Vale sounded weary. He looked weary, too—his hair unkempt, his eyes shadowed, like he'd gotten very little rest or food.

"I didn't know if you would survive," he said quietly. "You were very sick."

Most don't survive the process, he had told me.

The process.

Only now did it start to sink in, what had happened to me —what I had done. My human self had withered and died, just as it had always been destined to.

And I...

I rubbed my fingers together. Even my skin felt different. Smoother. Unmarked.

Gods. The shock left me dizzier than my illness. The words even sounded strange aloud.

"You Turned me."

Vale nodded slowly. Hesitantly.

"I asked you—"

"I said yes."

I want to stay.

And so, he'd helped me stay.

"Yes," he whispered.

I met his eyes. He didn't blink, watching me carefully—as if to make sure he saw every shade of my reaction to this.

"I won't lie to you, mouse. It won't be an easy transition. A part of you did die that day. A different version of you was born. There will be things you'll grieve. There will be things about yourself you'll need to learn how to embrace. Things that might be... uncomfortable. But..."

His hand fell over mine as his voice faded. He cleared his throat a little. "But you'll have help."

I took this in for a long moment.

He asked quietly, "Do you regret it?"

Regret it?

I felt... different. So wildly different than I always had in every way, shedding not only my humanity, but the ever-present looming threat of time.

Even through my illness, I felt the strength lying in wait, ready to be seized. This body wouldn't wither. It would thrive.

But I couldn't care less about that.

The prospect that overwhelmed me was the thought of *time*.

Time. So much of it. Time to collect knowledge. Time to

see the world. I didn't know what I might do with so much of it.

I felt strange, yes. I could already tell Vale was right that it would take me a long time to adjust to this new existence.

But regret?

"No," I said. "I don't."

Vale's shoulders lowered slightly, as if in relief. He avoided my gaze, rolling my fingers gently through his. My senses were so heightened, I could feel every wrinkle and texture of his skin.

"You... you came back," I said.

"I know it wasn't what you wanted me to do. But I was a general because I was better at giving orders than following them."

Not true. I wanted it more than anything. For him to come back. Even if I didn't know it at the time.

"Why?" I asked.

"You were right. The roses were special."

I smiled a little. "You finally noticed."

"They never died."

They look exactly the same as they always have, he'd said, so irritated, like I'd tricked him. I'd thought it was funny at the time. Of course a vampire wouldn't notice the absence of decay, the absence of time, when they lived beyond it themselves.

"When I was preparing to leave," he said, "I was gathering the roses. And I noticed, when I held them, that one of them had begun to wither—just a little. I've held god-touched objects before. And when I was touching them, I—I felt it. It feels strange, for us to touch an object touched by the White Pantheon."

Us.

Him and I. Vampires.

174

But that struck me less than the image of what he was describing. That Vale, when packing up his belongings, had not only taken the roses with him, but had sat there holding them. For a moment I could picture it so vividly, him cradling those roses, and it made my chest tighten.

His thumb rubbed the back of my hand.

"It was foolish that I didn't realize you were god-touched, too. You strange creature." A wry smile tugged at his lips. "Different from any human or any vampire I had ever encountered."

Gods, the way he looked at me—a strange feeling shivered in my heart.

But then my brow furrowed.

"But how did you *know?*" I said.

Vale had pieces of the truth. Incomplete evidence. But not enough to draw a final conclusion.

He lifted one shoulder in an almost-shrug. "I didn't know, Lilith. I felt."

So few words, and yet they encapsulated something I had struggled to name in those final moments. Something I understood, against all reason and logic.

"I knew that—that I would be making a mistake, in leaving you," he said softly. "I knew it, even if I couldn't name precisely why. So I came for you."

And he had saved me.

My throat thickened. I swallowed, though it was difficult through the dryness of my throat.

"And what about Adcova?"

"Ah, the best part." He smoothed my hair from my face. He'd been doing that this whole time—touching me in all these little mundane, fussing ways. Smoothing hair, adjusting my sleeve, wiping beads of sweat. "It seems," he said, "that Adcova has escaped its god's ire at last."

I let out a rough exhale. I almost didn't want to believe it. Didn't want to hope it could be true.

"I asked my errand boy to send updates," he went on. "There have been no new cases reported in town, or anywhere else in the area. And it seems a peculiar new drug has cured the cases that already existed."

The pride shone in his voice. My chest hurt fiercely, a strange burning sensation. I couldn't speak. He held my hand tight.

"It's over, Lilith," he said. "You saved them."

Years. Years of my life. Countless hours in my study, countless hours of sleep stolen. Thousands of books, thousands of notes. Years-worth of pen-grip callouses on my fingers.

For this.

For...

"Mina," I choked out.

I'd meant for it to be a real question, but I couldn't get it out, not without breaking down.

Vale was silent for too long, making worry tighten in my stomach. He let go of my hand—somewhat reluctantly—and went to the door.

And when she appeared in the doorway, my heart cracked open.

She was bright and vivacious and full of life like I hadn't seen her in years, as if all those layers of death she had shed in the form of dusty skin on our floors had left her a whole new person. New, and yet, the version of her I had always known.

She smiled at me through tears, a huge, sun-bright grin, and I opened my mouth to speak and let out a garbled sob.

She crossed the room in several clumsy rushed steps and threw herself against me in an embrace.

"I know," she said, when I couldn't speak, and neither of us said anything else.

Because for so long, I had struggled to connect with my sister. Struggled to show her the warmth beneath my cold. Struggled to let her see the love my face and words couldn't convey to her.

I'd thought I'd die with her thinking I did not love her.

I did die, and that fear died with me.

Because here, in this moment, with me on the right side of death and her on the right side of living, lost in a tearful embrace hello instead of goodbye, we met each other on level ground.

Here, we understood each other so completely, words were useless, anyway.

WE DID, eventually, let each other go and compose ourselves, and I did, eventually, manage to get my grip on words again.

We made it through a few awkward minutes of stilted conversation before the question I couldn't help but ask bubbled to the surface.

"Do you hate me?" I asked. "Or hate... what I've become?"

Mina's eyes widened. Her answer was immediate. "Never. I could never hate you, Lilith."

"Do I look different now?"

I was curious, I had to admit. There was no mirror in this room, and I definitely wasn't strong enough to get up and go look for one.

She thought about this before answering.

"You look different," she said, "but you also look more like yourself than you ever had. And that makes sense, because you were never... like us. You were always so different than the rest of us."

She said it with such warmth, even though I'd always resented my differences from those around me.

"You'll be going with him," she said. "Right? To Obitraes."

I hadn't even been able to think that far ahead yet. I touched my throbbing temple.

"He hasn't asked me to."

Not technically true. He did ask me to go with him—a lifetime ago, before I went to Vitarus.

Mina gave me a flat stare. "He'll ask."

"I don't have to. I could live outside the town." It was risky, and the last thing I would want to do is draw more negative attention to Adcova. But Vale had managed it for centuries. Maybe I could.

She looked at me like I was insane.

"Why would you do that?"

"Because..."

I had never been farther than twenty miles away from my home. I had a sister who had always needed me, a cause that demanded all my focus and energy.

"That would be a stupid thing to do," she said, so plainly I almost laughed. "I'm not as smart as you, but I'm no idiot. You think I don't know what you want? I know you've always wanted to travel. See new things. Learn new things. So go!"

She smiled, even though her eyes were damp again. She took my hand and squeezed. "You've spent your whole damned life dying, Lilith. Now you've gotten that out of the way, and you get to go live."

I was silent, a bit struck.

My voice was rough when I said, "You know that I never wanted to leave you."

I didn't mean here, in this moment.

I meant all those days when she asked me to stay, and I went to my office instead.

I meant all those years when she, and my parents, and my friends, and everyone around me begged for me to stay, when death was stealing me away instead.

Her face softened.

"I know," she said. "Of course, I know."

She said it like it was obvious and simple, and a silly thing that didn't need to be clarified.

I always had thought that Mina didn't understand me, all my true intentions hidden behind the wall I couldn't figure out how to scale between me and the people around me.

Maybe she did see more than I ever realized, after all.

CHAPTER TWENTY-SIX

My strength came back slowly. Vale warned me that recovery would be long and challenging. I wouldn't have my full strength for months longer. He didn't know that I already felt better than I had for most of my life.

Mina departed a week after I woke. She had a life to get back to, after all. Letting her go was bittersweet. I didn't cry this time—I'd gotten all that out of my system when we first reunited—but I did watch her leave in the moonlight, a lump in my throat.

Later that night, Vale came into my room. While Mina was here, he slept in a different room—or, if I was feeling especially unwell, stayed in the armchair of mine. Now, with her gone, we both seemed acutely aware of our sudden privacy.

He stood awkwardly in the door, fussing with random items on the bureau.

I watched him, a smile tightening my cheeks, feeling

strangely warm in a way that I suspected had nothing to do with my lingering fever.

"I have something I've been meaning to give you," I said.

He turned, eyebrow twitching. "Oh?"

"Well, something I promised you."

I leaned to the bedside table and reached into the drawer. Vale sat at the edge of the bed.

"Put out your hands."

He did, and within them, I placed one last rose.

Mina had given it to me shortly before she left. "I thought you might want to remember home," she had said with a wink, somewhat drily—a dark joke about all we'd escaped.

I had to drop the flower into his cupped palm, because it was falling apart. The stem was a dried-out stalk, the leaves crumbling, the petals disintegrating in dusty patches of black and faded dark red.

Vale let out a low chuckle.

"And to think I was going to forgive your debt."

"I promised you six roses. Thus, I have provided six roses."

"And I have given you plenty of my blood," he said. "So it looks like our deal is done."

He smiled at me in a way that was, perhaps, supposed to be coy, but felt sadder—more uncertain. It occurred to me that it had probably been a very long time since Vale had to deal with uncertainty.

I was still feverish and dizzy, but I leaned across the bed, turned his face towards me, and kissed him.

His arms folded around me, pulling me closer to him. He deepened the kiss immediately, like he'd been waiting for it, and it consumed me—with my senses heightened like this, I lost myself in his taste, his texture, the way his breath quickened a little when the full length of my body pressed against his. I wound my arms around his neck and reacquainted myself with him, running my hands over his shoulders, his back, his throat, his hair.

And he did the same, touching me like all this time he hadn't been completely sure that I was really here. Not until now, when he had to reaffirm every angle of my form.

I pulled away just barely, just enough to tilt my head to a new angle.

But he said, before I could move, "I need to ask you something."

I paused there, so close our noses were touching. My eyes flicked up to meet his. Our clothes were still on, our mouths parted, and yet I felt so staggeringly connected to him in this moment—our breaths nearly matching in cadence, slightly unraveled, shared between us.

"Ask," I whispered.

"I need to return to Obitraes." He leaned a little closer, so each word brushed my mouth. And here he snuck a little kiss in, a faint graze of his tongue over my lips. "I intend to go back to the House of Night."

"And?"

Another touch, barely a kiss—mine, this time.

"Come with me," he said.

He exhaled the plea, and I took it into my lungs.

"Yes," I said, giving him my answer in my next breath.

Our next kiss was longer, deeper. I melted against him. The next thing I became aware of, I was against the bed, Vale lying beside me.

He pulled away.

"You have to understand what you're agreeing to. It's a nation at war. I'm not sure what we'll be returning to."

He held my shoulder, firmly. And though his gaze kept wandering down to my mouth, it always came back to my eyes—examining me, making sure he understood the answer to this important equation.

The answer to the equation of what I would do now, with my new, endless life.

The easiest question in the world.

"Do I seem, Lord Vale," I said, "like someone frightened of the unknown?"

His eyes crinkled with a smile.

"Nosy mouse," he murmured, and this time, when I tried to kiss him, he let me. We wound ourselves around each other. My thighs opened around him. I gave him every one of my new, heightened senses, and for the first time in my life, I felt so utterly at ease with the world that surrounded me.

Vale had discarded the withered rose in favor of my skin. The petals spread around us, now nothing but decaying dust.

Unnatural life.

Rightful death.

And Vale and I, between both, beholden to neither, and everything we were ever meant to be.

READY FOR MORE?

Thank you for reading Six Scorched Roses — I hope you loved it! If you're ready for more epic vampire adventures in this world, you should pick up *The Serpent and the Wings of Night*.

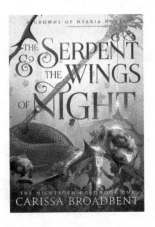

Human or vampire, the rules of survival are the same: never trust, never yield, and always — *always* — guard your heart.

The adopted human daughter of the vampire Nightborn king, Oraya has learned to survive in a world designed to kill her. But her only chance at becoming more than prey is winning the Kejari, a legendary tournament held by the goddess of death. To survive, she's forced to ally with a mysterious rival — a vampire who's her greatest competition, an enemy to her father's throne, and far, far too interested in her.

Get The Serpent and the Wings of Night now — free in KU!

AUTHOR'S NOTE

Thank you so much for reading Six Scorched Roses! This little book was a bit of a strange beast. When I originally wrote it for the Flirting with Darkness anthology, I honestly wasn't sure how people would react to it. If you've read The Serpent and the Wings of Night, you know that this is a wildly different story—slower and more romantic rather than action-packed and bloodthirsty. Plus, it's the shortest book I've ever written. While I loved writing it, I was also honestly prepared for readers to have some very conflicted feelings about it!

Instead, readers connected more with Vale and Lilith— especially Lilith!—than I ever thought they would. I loved seeing how many people saw themselves in her and her struggle to connect meaningfully with other people. If you've just read it for the first time, I hope that you loved it, and that perhaps you, too, found something relatable in Lilith's story.

Six Scorched Roses is a part of the world of the Crowns of Nyaxia series, and Vale and Lilith play significant roles in the world from book 2 onwards. If you're interested in following more about them, you might want to pick up The Serpent and the Wings of Night and dive deeper into this dark, vampiric world!

If you enjoyed this book, I would truly appreciate if you'd consider leaving a review on Amazon or GoodReads.

And if you'd like to be the first to know about new releases, new art, new swag, and all kinds of other fun stuff, consider signing up for my newsletter at carissabroadbent books.com, hanging out in my Facebook group (Carissa Broadbent's Lost Hearts), or joining my Discord server (invite at linktr.ee/carissanasyra!).

I hope we keep in touch!

ALSO BY CARISSA BROADBENT

The Crowns of Nyaxia Series

Book 1: The Serpent and the Wings of Night

Book 2: The Ashes and the Star-Cursed King

Book 3-6: Coming soon!

Related Standalone Novella: Six Scorched Roses

The War of Lost Hearts Trilogy

Book 1: Daughter of No Worlds

Book 2: Children of Fallen Gods

Book 3: Mother of Death and Dawn

GLOSSARY OF TERMS OF THE NYAXIA WORLD

ACAEJA - The goddess of spellcasting, mystery, and lost things. Member of the White Pantheon.

ADCOVA - A small farmers' town located in the continent of Dhera. Lilith's home. Years ago, it was cursed by the god Vitarus with a terrible, mysterious plague that has worsened ever since.

ALARUS - The god of death and husband of Nyaxia. Exiled by the White Pantheon as punishment for his forbidden relationship with Nyaxia. Considered to be deceased.

ATROXUS - The god of the sun and leader of the White Pantheon.

BLOODBORN - Vampires of the House of Blood.

BORN - A term used to describe vampires who are born via

biological procreation. This is the most common way that vampires are created.

DHERA - A continent in the human lands. Vale and Lilith are currently living there.

OBITRAES - The land of Nyaxia, consisting of three kingdoms: The House of Night, The House of Shadow, and the House of Blood.

HEIR MARK - A permanent mark that appears on the Heir of the Hiaj and Rishan clans when the previous Heir dies, marking their position and power.

HIAJ - One of the two clans of Nightborn vampires. They have featherless wings that resemble those of bats.

THE HOUSE OF BLOOD - One of the three vampire kingdoms of Obitraes. Two thousand years ago, when Nyaxia created vampires, the House of Blood was her favorite House. She thought long and hard about which gift to give them, while the Bloodborn watched their brothers to the west and north flaunt their powers. Eventually, the Bloodborn turned on Nyaxia, certain that she had abandoned them. In punishment, Nyaxia cursed them. The House of Blood is now looked down upon by the other two houses. People from the House of Blood are called **BLOODBORN**.

THE HOUSE OF NIGHT - One of the three vampire kingdoms of Obitraes. Known for their skill in battle and for their vicious natures, and wielders of magic derived from the night sky. There are two clans of Nightborn vampires, **HIAJ** and

RISHAN, who have fought for thousands of years over rule. Those of the House of Night are called NIGHTBORN.

THE HOUSE OF SHADOW - One of the three vampire kingdoms of Obitraes. Known for their commitment to knowledge; wielders of mind magic, shadow magic, and necromancy. Those of the House of Shadow are called SHADOWBORN.

IX - Goddess of sex, fertility, childbirth, and procreation. Member of the White Pantheon.

KAJMAR - God of art, seduction, beauty, and deceit. Member of the White Pantheon.

NECULAI VASARUS - The former Rishan king of the House of Night. Usurped and killed 200 years prior to the events of this book.

NIGHTBORN - Vampires of the House of Night.

NIGHTFIRE - Like Asteris, another form of star-derived magic wielded by the vampires of the House of Night. While Asteris is dark and cold, Nightfire is bright and hot. Nightfire is commonly used in the House of Night but very difficult to wield masterfully.

NYAXIA - Exiled goddess, mother of vampires, and widow of the god of death. Nyaxia lords over the domain of night, shadow, and blood, and as well as the inherited domain of death from her deceased husband. Formerly a lesser goddess, she fell in love with Alarus and married him despite the forbidden nature of their relationship. When Alarus was murdered by the White Pantheon as punishment for his marriage to her, Nyaxia broke free from the White Pantheon in a fit of rage, and offered her supporters the gift of immortality in the form of vampirism—founding Obitraes and the vampire kingdoms. *(Also referred to as: the Mother; the Goddess; Mother of the Ravenous Dark; Mother of Night, Shadow, and Blood)*

RISHAN - One of the two clans of Nightborn vampires. Have feathered wings. Usurped by the Hiaj 200 years ago.

SALINAE - A major city in the House of Night. Located in Rishan territory. When the Rishan were in power, Salinae was a thriving hub, functioning as a second capital. Oraya spent the first years of her life, before Vincent found her, in Salinae.

SHADOWBORN - Vampires of the House of Shadow.

SIVRINAJ - The capital of the House of Night. Home to the Nightborn castle, the Moon Palace, and host to the Kejari once every 100 years.

SRANA - The goddess of knowing and science. Member of the White Pantheon.

TURNING - A process to make a human into a vampire, requiring a vampire to drink from a human and offer their blood to the human in return. Vampires who underwent this process are referred to as **TURNED**.

VITARUS - The god of abundance and famine. Member of the White Pantheon.

WHITE PANTHEON - The twelve gods of the core cannon, including Alarus, who is presumed deceased. The White

Pantheon is worshipped by all humans, with certain regions potentially having favor towards specific gods within the Pantheon. Nyaxia is not a member of the White Pantheon and is actively hostile to them. The White Pantheon imprisoned and later executed Alarus, the God of Death, as punishment for his unlawful marriage with Nyaxia, then a lesser goddess.

Zarux - The god of the sea, rain, weather, storms, and water. Member of the White Pantheon.

ACKNOWLEDGMENTS

I've written a number of these at this point, and it never gets less surreal! I have so many people to thank not only for helping me bring this particular story into the world, but also for helping me build the career I feel so fortunate to have today.

First on the list will always be Nathan, the love of my life, my best friend, my number one brainstorming buddy and supporter, and the source of all my free art (haha, thanks). I could do none of this without you and could never ask for a better partner.

May Sage, thank you for inviting me to the anthology that originally kicked off this adventure! I loved working with you and having the opportunity to write this little story.

Clare Sager, thank you for being the best coworker, author wife, friend, and human ever. Thank you for the many many many hours spent listening to the Carissa-brain-to-mouth pipeline. I couldn't have written this book, let alone survive this business at all, without you!

K.D. Ritchie, thank you so much for this *incredible* cover design. Seriously, you truly outdid yourself on this one! I absolutely love working with you.

Noah, thank you for your always-impeccable editing skills and for doing so on such a bizarro timeline and delivery schedule. You are the best!

Anthony, thank you for a fabulous set of proofreading eyes.

Rachel, thank you for *another* fabulous set of proofreading eyes... and for all the hilarious comments! I really love them!

And finally, as always, I'm saving the best for last: thank you to *you*, lovely reader, for coming along on this ride with me. Whether this is your first foray into my books or you've been around for quite awhile, I hope you loved it and I am so so grateful to have you. None of this would exist without you. Love you!

ABOUT THE AUTHOR

Carissa Broadbent has been concerning teachers and parents with mercilessly grim tales since she was roughly nine years old. Since then, her stories have gotten (slightly) less depressing and (hopefully a lot?) more readable. Today, she writes fantasy novels with a heaping dose of badass ladies and a big pinch of romance. She lives with her husband, one very well behaved rabbit, one very poorly behaved rabbit, and one perpetually skeptical cat in Rhode Island.

instagram.com/carissabroadbentbooks

tiktok.com/carissabroadbent

facebook.com/carissabroadbentbooks

twitter.com/carissanasyra